FLA. SOLAR ENERGY CENTER LIBRARY

Fred Schäfer and Richard van Basshuysen

Reduced Emissions and Fuel Consumption in Automobile Engines

Springer-Verlag
Wien New York

LIBRARY
Florida Solar Energy Center
1679 Clearlake Road
Cocoa FL 32922-5703

Univ.-Prof. Dr. Ing. Fred Schäfer
Fachhochschule Iserlohn, Iserlohn,
Federal Republic of Germany

Dipl.-Ing. Richard van Basshuysen
Bad Wimpfen, Federal Republic of Germany

Original title: "Schadstoffreduzierung und
Kraftstoffverbrauch von Pkw-Verbrennungsmotoren
(Die Verbrennungskraftmaschine. Neue Folge,
Band 7)", translated by J. Stähle
© 1993 Springer-Verlag/Wien

This work is published simultaneously by Springer-Verlag Wien New York and by Society of Automotive Engineers, Inc. and is subject to copyright.
All rights are reserved, whether the whole or part of the material is concerned specifically those of translation, reprinting, re-use of illustrations, broadcasting reproduction by photocopying machines or similar means, and storage in data banks.

© 1995 Springer-Verlag/Wien

Printed in Germany

Sole distribution rights for North America: Society of Automotive Engineers, Inc., Warrendale, PA

Data conversion: Thomson Press (India) Ltd., New Delhi
Printing and Binding: Druckerei zu Altenburg GmbH, D-04600 Altenburg

Printed on acid-free and chlorine-free bleached paper

With 275 Figures

```
Library of Congress Cataloging-in-Publication Data


Schäfer, Fred, 1948-
    [Schadstoffreduzierung und Kraftstoffverbrauch von PkW
 -Verbrennungsmotoren. English]
    Reduced emissions and fuel consumption in automobile engines/ Fred
 Schäfer and Richard van Basshuysen.
        p.   cm.
    Translation of: Schadstoffreduzierung und Kraftstoffverbrauch von
 PkW-Verbrennungsmotoren.
    Includes bibliographical references.
    ISBN 3-211-82718-8 (alk. paper)
    1. Motor vehicles--Pollution control devices.  2. Motor vehicles-
 -Fuel consumption.  3. Motor vehicles--Motors--Exhaust gas-
 -Environmental aspects.  4. Air--Pollution--Law and legislation.
 I. Van Basshuysen, Richard, 1932-   .  II. Title.
 TL214.P6S35   1995
 629.25'04--dc20                                              95-38356
                                                                  CIP
```

ISBN 3-211-82718-8 Springer-Verlag Wien New York
ISBN 1-56091-681-8 Society of Automotive Engineers, Inc.

Preface

The enormous increase in environmental pollution caused by combustion processes makes it essential that solutions are found for their reduction. A major portion of this problem is caused by the exhaust emissions from combustion engines.

In the early 1960's, the first laws governing emissions were passed in the USA. The introduction of the λ-regulated three-way catalytic converter was a breakthrough in the reduction of pollutant emissions in Europe and Japan as well.

Over the last several years, there has been much discussion on the interrelation of CO_2 emissions with the global warming phenomenon. This in turn has increased pressure to develop and produce more fuel efficient engines and vehicles.

This is the central topic of this book. It covers the underlying processes which cause pollutant emissions and the possibilities of reducing them, as well as the fuel consumption of gasoline and diesel engines, including direct injection diesel engines. As well as the engine-related causes of pollution, which are found in the raw exhaust, there is also a description of systems and methods for after-treatment of emissions. The significant influence of fuels and lubricants (both conventional and alternative fuels) on emission behavior is also covered. In addition to the conventional gasoline and diesel engines, lean-burn and direct injection gasoline engines and two-stroke gasoline and diesel engines are included. The potential for reducing fuel consumption and pollution is described, as well as the related reduction of CO_2 emissions. Finally, a detailed summary of the most important laws and regulations pertaining to pollutant emissions and consumption limits is presented.

This summary covers the actual laws in the United States, including California, as well as Europe and Japan. Additionally, the different international testing methods are described, together with a description of the most important emission measuring instruments and methods.

This book is intended for practising engineers involved in research and applied science as well as for interested engineering students.

The authors would like to thank the following people for their generous advice and numerous suggestions: Dr. Cartellieri, AVL; Dr. Fraidl, AVL, Dr. Laurenz, Audi; Prof. Dr. Jordan, University of Cologne; Prof. Dr. Schroeder, University of Hamburg.

We also wish to thank the following companies for providing charts and graphs: Audi, AVL, Bosch, FEV, Ford, Mercedez-Benz, Opel (GM), Pierburg, Porsche, Shell, TÜV-Rheinland and VW.

Prof. Dr. Fred Schäfer   Iserlohn, July 1995
Dipl.-Ing. Richard van Basshuysen   Bad Wimpfen, July 1995

Contents

List of symbols and abbreviations		IX
1	Introduction	1
2	Causes of pollutants and their toxicity	3
2.1	Combustion processes	3
2.1.1	Chemical reactions	3
2.1.2	Combustion products	4
2.2	Toxicity and effects of pollutants on mankind	6
2.3	Causes of pollutants in engines	7
2.3.1	Causes of pollutants in gasoline engines	7
2.3.1.1	Pollutant components	8
2.3.2	Causes of pollutants in diesel engines	10
2.3.2.1	Pollutant components	11
2.4	Causes and effects of pollutants in the atmosphere	13
2.4.1	CO_2 and the climate	18
3	Design features which influence pollutant emissions and fuel consumption in four-stroke engines	20
3.1	Gasoline engines	21
3.1.1	External mixture preparation	21
3.1.1.1	Combustion chamber shape, layout and compression ratio	21
3.1.1.2	Mixture preparation, mixture control and direct fuel injection systems	24
3.1.1.3	Ignition timing and spark plugs	26
3.1.1.4	Exhaust gas recirculation (EGR)	28
3.1.1.5	Valve timing	29
3.1.1.6	Inlet port configuration and swirl	32
3.1.1.7	Bore-stroke relation and cylinder volume	34
3.1.1.8	Cooling system	35
3.1.2	Lean burn engine concepts	37
3.1.2.1	Lean burn management systems	41
3.1.3	Internal mixture preparation	42
3.1.4	Gasoline engine potential in terms of emissions and fuel consumption	47
3.2	Diesel engines	47
3.2.1	Combustion process	48
3.2.1.1	Pre-chamber engines	49
3.2.1.2	Direct injection diesel engines	54
3.2.2	Compression ratio	64
3.2.3	Fuel injection hydraulics	64
3.2.4	Electronic control of diesels	67
3.2.5	Exhaust recirculation	69
3.2.6	Diesel engine potential in terms of emissions and fuel consumption	69
3.3	Characteristic engine emission maps	71
3.4	Long-term emission stability	72
4	Engine-related measures which reduce pollutant emissions and fuel consumption in two-strokes	77
4.1	Gasoline engines	80
4.2	Diesel engines	82
5	Exhaust aftertreatment methods	86
5.1	Gasoline engines	86
5.1.1	The effect of post-combustion reactions in exhaust systems	86
5.1.2	Thermic reactors	86
5.1.3	Catalytic converter systems	88
5.1.3.1	Working principles of catalytic converters	89
5.1.3.2	Catalytic converter design	90
5.1.3.3	Conversion efficiency	93
5.1.3.4	Location in the vehicle	98
5.1.3.5	Disadvantages	98
5.1.3.6	Critical operating range	99
5.1.3.7	Recycling	99
5.1.3.8	Lambda sensors	100
5.2	Diesel engines	102

15.2.1	Thermic reactors	102
5.2.2	Catalytic reactors	103
5.2.2.1	Oxidation catalytic converter	104
5.2.3	Separation systems	106
5.2.3.1	Trap systems	107
5.2.3.2	Trap regeneration	110
6	The influence of fuel and lubricants on emissions and fuel consumption	117
6.1	Conventional fuels	117
6.1.1	Gasoline	117
6.1.1.1	Reformulated gasoline	119
6.1.2	Diesel fuels	119
6.1.2.1	Reformulated diesel fuel	122
6.2	Alternative fuels	122
6.2.1	Alternative fuels for spark-ignited engines	124
6.2.1.1	Alcohols (methanol and ethanol)	127
6.2.1.2	Gases	128
6.2.1.3	Hydrogen	129
6.2.2	Alternative fuels for compression ignition engines	132
6.2.2.1	Alcohols (methanol and ethanol)	132
6.2.2.2	Vegetable oils	133
6.2.2.3	Rape seed methyl ester (RME)	134
6.2.2.4	Other fuels	135
6.3	Effects of fuel additives	136
6.3.1	Additives for gasoline engines	137
6.3.2	Additives for diesel engines	138
6.4	Effects of lubricants	139
7	Problems with CO_2 emissions	142
7.1	CO_2 emissions and their causes	142
7.2	CO_2 emissions from motor vehicles	142
7.3	CO_2 emissions and fuels	144
8	Laws regulating the emission of pollutants and maximum fuel consumption of combustion engines (as of 1992)	147
8.1	Testing procedures	147
8.2	Measurement methods and instruments	150
8.2.1	Emission measurement methods	150
8.2.2	Emission measuring instruments	152
8.3	Pollutant limits and maximum fuel comsumption values	156
8.3.1	USA – 49 states	156
8.3.2	USA – California	157
8.3.3	Japan	157
8.3.4	European Economic Community	162
8.3.5	Federal Republic of Germany	162
8.4	Outlook on future developments in emissions and consumption laws	164
8.4.1	USA – 49 states	164
8.4.1.1	Emission limits	164
8.4.1.2	Testing methods	165
8.4.1.3	Maximum fuel consumption and other regulations	166
8.4.2	USA – California	167
8.4.2.1	Emission limits	167
8.4.2.2	Testing methods	171
8.4.2.3	Maximum fuel consumption and other regulations	171
8.4.3	Japan	172
8.4.3.1	Emission limits	172
8.4.3.2	Testing methods	172
8.4.3.3	Maximum fuel consumption and other regulations	173
8.4.4	European Economic Community	174
8.4.4.1	Emission limits	174
8.4.4.2	Testing methods	176
8.4.4.3	Maximum fuel consumption and other limits	177
8.4.5	Federal Republic of Germany	178
8.4.5.1	Emission limits	178
8.4.5.2	Testing methods	178
8.4.5.3	Maximum fuel consumption and other limits	178
References		182
Subject index		191

List of symbols and abbreviations

A	Pre-exponential coefficient
A_i	Elementary reaction involving component i
ASU	Annual exhaust test (Germany)
EGR	Exhaust gas recirculation
ATL	Exhaust turbocharger
b_e	Specific fuel consumption [g/kWh]
CAFE	U.S. fuel consumption specifications
CARB	Californian Air Resources Board
CO	Carbon monoxide
CO_2	Carbon dioxide
CVS	Constant Volume Sampling
c_u, c_a	Circumferential speed, axial speed [m/sec]
CN	Cetane number (Cetane rating)
CNG	Compressed Natural Gas
DI	Direct injection (Direct-injection engine)
D	Swirl number
E	Activation energy
ECE	Economic Commission for Europe
EB	Start of inlet
EEC	European Community
EDR	Electronic Diesel control
EPA	Environmental Protection Agency (U.S.A.)
EUDC	New European extra-urban driving cycle
FE	Fuel Economy
FTP	Federal Test Procedure
FID	Flame ionization detector
HC	Hydrocarbon (Unburned hydrocarbons)
H/C	Ratio of hydrogen to carbon atoms in fuel
h	Valve lift [mm]
hwy	Highway
h.v	Radiation energy
i	Component of matter
IDI	Indirect injection (Pre-chamber engines)
lbs	Pound ($= 0.4536$ kgs)
j	Elementary reaction
J_j	Reaction velocity
k_j	Constant of reaction velocity
LA-4 cycle	Emission testing cycle
LPG	Liquefied Petroleum Gas
M90	Gasoline-methanol mixture (90% methanol)
MY	Model Year
M_d	Torque [Nm]
mpg	Miles per gallon
m_{H2}	Mass of hydrogen [kg]
m_{fu}	Mass of fuel [kg]
m_{fu}	Fuel input [kg/s]
m_L	Air mass [kg]
$m_{L\,stoich}$	Stoichiometric air mass
NMHC	Non-methane hydrocarbon
n	Engine speed [rpm]
NO_x	Nitrogen oxide
OBD	On-Board Diagnosis
OBVR	On Board Vapour Recovery
TDC	Top dead center
p	Pressure [bar]
Pme	Mean effective pressure [bar]
ppm	Parts per million

PM	Particulates
r_i	Concentration of component i [ppm] [% by volume]
RON	Research Octane Number
RME	Rape seed methyl ester
SN	Smoke number
R_m	Molar gas constant
S/D	Stroke-to-bore ratio
SD	Injection duration
SB	Start of injection
SHED	Test to determine evaporative losses
SI	Spark-ignition engine
SO_2	Sulphur dioxide
V	Volume
w_e	Effective specific work [kJ/dm^3]
w_i	Internal specific work [kJ/dm^3]
WK	Swirl chamber
ZV	Ignition lag
ZZ	Injection timing
Cyl	Cylinder
°CA	Degrees of crank angle
2V/4V	Two-valve, four-valve engine (valves per cylinder)
	Efficiency
λ	Air/fuel ratio
ε	Compression ratio
σ_i	Specific molecular number
$v''_{i,j}$	Stoichiometric coefficients of component i in reaction j

1 Introduction

The enormous increase in the use of fossil energy sources in the industrialized nations, in particular, has caused increased pollution of the air, with vehicle exhaust emissions being a significant contributing factor. By the late 50s, initial legal standards to limit emissions of major toxic exhaust components were therefore enforced in the U.S.A. The necessity of implementing such measures that have since gained importance in virtually all industrial nations is highlighted by the forecast annual mileage in Germany shown as an example in *Fig. 1.1* [1.1].

The development in Germany, broken down by passenger vehicles with Diesel and spark ignition (SI) engines, is shown in *Fig. 1.2*. The annual mileage was calculated from demographic data [1.2].

The share of diesel-engined passenger vehicles will continue to increase slightly for some time whereas the share of passenger vehicles without catalytic converters will decrease rapidly. The diagram shows, however, that a significant share of SI-engined passenger cars without catalytic converters will remain in use well beyond the year 2000. This may be attributed to the long life cycle of a vehicle of approx. 20 years after its production introduction (production time approx. 7 years, vehicle life approx. 13 years).

The maximum annual mileage of passenger vehicles will probably be reached around the turn of the millenium.

This is due to a saturation of the market and the development of the population structure. The major share will be accounted for by passenger vehicle engines with displace-

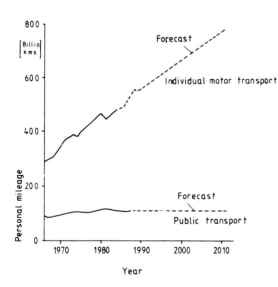

Fig. 1.1 Development of annual mileage [1.2]

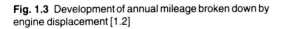

Fig. 1.3 Development of annual mileage broken down by engine displacement [1.2]

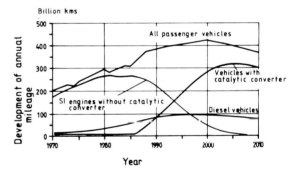

Fig. 1.2 Breakdown of annual mileage of passenger vehicles [1.2]

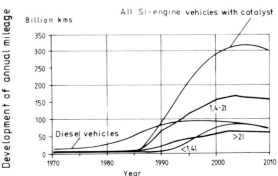

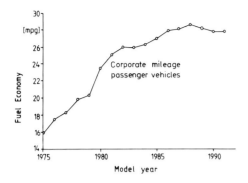

Fig. 1.4 Development of fuel consumption [1.6]

Fig. 1.5 Fuel consumption of prototype vehicles [1.6]

Company	Power [KW]	Fuel Economy [mpg]	Model
General Motors	28	61 city 74 hwy	Otto
British Leyland	53	41 city 52 hwy	Otto
Volkswagen	39	63 city 71 hwy	Diesel
Volkswagen	38	74 city 99 hwy	Diesel
Volvo	38, 65	63 city 81 hwy	Diesel
Renault	37	63 city 81 hwy	Diesel
Renault	20	78 city 107 hwy	Otto
Peugeot	37	55 city 87 hwy	Diesel
Peugeot	21	70 city 77 hwy	Otto
Ford	29	57 city 92 hwy	Diesel
Toyota	41	89 city 110 hwy	Diesel

ments from 1.4 to 2.0 liters. *Figure 1.3* shows the share of passenger cars with diesel engines as well as the breakdown of catalytic-converter passenger cars by displacement [1.2].

Pollutant emissions will decrease only slowly. Contributing factors in this context are a continuing increase of average mileage as well as the relatively slow reduction of the number of non-emission control vehicles. These facts may be counteracted by more stringent emission legislation and, on a worldwide basis, by extending emission legislation to other nations. As far as fuel requirements for the next century are concerned, supply bottlenecks may appear on the horizon. Even if we disregard such scenarios for the moment, however, additional drastic reductions of fuel consumption are required in order to cope with the CO_2 emissions problem closely related to the greenhouse effect [1.5].

Figure 1.4 shows the development of fuel consumption in the U.S.A. from 1975 to 1991. In recent years, the rate of reduction of fuel consumption has slowed down considerably and now stands at a corporate average figure of slightly below 28 mpg for all newly-registered vehicles.

To allow the significant potential for reducing fuel consumption to be exploited, suitable measures applicable to engines and vehicles should be introduced. According to [1.6], weight reduction, new transmission concepts and improved aerodynamics are considered to be significant vehicle-related measures.

Reduced fuel consumption of spark-ignition engines may be accomplished by using variable valve trains, multi-valve technology, improved fuel injection systems, direct injection, cylinder cutout, reducing friction and increasing compression ratio. In the future, lean-burn engines, lean mix engines, two-stroke engines and lean-burn catalytic converters (NO_x) will offer an additional potential for reducing fuel consumption. At the same time, all the above measures help to reduce emissions of the greenhouse gas CO_2 that affects the climate on a worldwide scale.

For the diesel engine, further development of the direct-injection engine is a key issue.

The forecast fuel consumption figures to be expected in the future vary widely. As an example, the corporate U.S. average fuel economy for the 1995/96 Model Year (MY) is specified at 28 to 34 mpg, at 28.7 to 45 mpg for MY 2000/2001, and at 45 to 74 mpg for MY 2010 [1.6]. *Figure 1.5*, which shows some particularly economical prototype vehicles, indicates that the quoted figures are actually realistic [1.7].

Vehicles with a fuel consumption of three liters per 100 km will become a reality [1.3] [1.4].

2 Causes of pollutants and their toxicity

2.1 Combustion processes

Any combustion process in the engine involves oxidation with oxygen. Normally engines use fuels that consist of hydrocarbons. Both gasoline and diesel fuel are made up of mixtures that include a large number of different hydrocarbons.

2.1.1 Chemical reactions

When a hydrocarbon molecule is burned "completely", this process, in theory, only releases carbon dioxide and water.

As an example, methane oxidation involves the following substances:

$$CH_4 + 2O_2 \Leftrightarrow CO_2 + 2H_2O.$$

Methane is a gas and is dissolved in small quantities in spark-ignition engine gasoline. Some hydrocarbons contain additional oxygen in the form of OH groups, e.g. methanol (CH_3OH). Fuel composition therefore has a significant effect on exhaust gas composition.

The oxidation process involves a number of intermediate products and intermediate reactions that have a more or less extended life.

The kinetics of homogeneous gas reactions may be used as an example to describe this change of state as a function of time under the impact of chemical reactions. Observations of kinetics allow the time-related characteristics of reactions and of the components involved to be assessed. The effect that each individual elementary reaction has on the formation and decomposition of a component, respectively, is taken into account in this process. The totality of the elementary reactions (see *Fig. 2.1*) is characteristic of the reaction scheme of the process investigated [2.35]. This reaction scheme is described by j elementary reactions in the form of

$$\sum_{i=1}^{n} v'_{i,j} \cdot A_i \Leftrightarrow \sum_{i=1}^{n} v''_{i,j} \cdot A_i.$$

The units $v'_{i,j}$ and $v''_{i,j}$ represent the stoichiometric coefficients of component i in reaction j.

The definition of the reaction velocity is used as a basis:

$$J_j = \frac{dn_i}{V(v''_{i,j} - v'_{i,j}) \cdot dt}.$$

This equation describes the quantity of matter of component i shifted per time and unit of volume in reaction j and is used to describe the change of an intensive variable of state vs time, e.g. the specific mol number σ_i of component i.

$$\frac{d\sigma_i}{dt} = \frac{1}{\rho} \cdot \sum_{j=1}^{r} (v''_{i,j} - v'_{i,j}) \cdot J_j$$

The reaction velocity J_j may be calculated by using the reaction velocity constants k_j

$$J_j = k_j \cdot \prod (\sigma \cdot \rho)^{v''_{i,j}}$$

These, in turn, are exponentially dependent on a variety of factors according to the following equation:

$$k_j = A \cdot T^n \cdot \exp(-E/R_m \cdot T).$$

The quantities $A, E/R_m$ and n are indicated in *Fig. 2.1*. A represents a coefficient used to characterize the number of surges that occur in a gas according to the theory of probability. E is the activation energy required for the reaction and n is an empiric temperature factor.

To illustrate the reaction of methanol with air, *Fig. 2.1* shows a possible reaction mechanism and the associated kinetic reaction data [2.1].

It is apparent that a large number of intermediate components are produced along with the CH_3OH, O_2, CO_2 and H_2O components. The quantity M in *Fig. 2.1* represents a surge partner that promotes energy exchange. The figures represent the reaction velocity and the activation energy of the reaction. The combustion process involving air generates additional nitrogen compounds. Nitrogen oxide compounds that are detrimental to the environment are also produced in this process. In the case of internal-combustion engine processes, the NO and NO_2 compounds are of particular importance.

For NO formation in the engine combustion process, the Zeldovic reactions are of importance. They are:

$$N_2 + O \Leftrightarrow NO + N,$$
$$O_2 + N \Leftrightarrow NO + O.$$

In the $\lambda > 1.2$ range, the reaction

$$OH + N \Leftrightarrow NO + H$$

constituting an extension of the Zeldovic mechanism becomes important [2.34] [2.2]. Along with the two reactions described, additional reactions producing NO_2 occur in this process.

In engine technology, the nitrogen oxides NO and NO_2 are usually combined and are referred to as NO_x. This unit is also used as a reference in legal regulations.

2.1.2 Combustion products

The reactions that occur in the engine combustion process not only produce numerous substances such as CO_2, H_2O, H_2, but also the following substances subject to legal limitations in the EEC and in Germany,

- Carbon monoxide (CO),
- Unburned hydrocarbons (HC),
- Nitrogen oxides (NO_X).

Other important pollutants contained in exhaust emissions are:

- Aldehydes (H–C–O compounds),
- Lead compounds produced by the use of leaded fuels,
- Sulphur dioxide (SO_2) of diesel fuels (due to use of fuel containing sulphur)
- Particulates including soot, especially with diesel engines.

As indicated before, CO_2 which has a particular effect on global warming of the atmosphere is also generated [2.3] [2.4] [2.5].

2.1 Combustion processes

No.	Reaction		$A[m^{3k}/k\,mol^k \cdot s]$	E/R_m	n
1	$CH_3OH + M$	$—CH_3 + OH + M$	$1,0 \cdot 10^{15}$	34200	0
2	$CH_3OH + CH_3$	$—CH_2OH + CH_4$	$1,8 \cdot 10^8$	4940	0
3	$CH_3OH + O$	$—CH_2OH + OH$	$1,7 \cdot 10^9$	1150	0
4	$CH_3OH + H$	$—CH_3 + H_2O$	$1,3 \cdot 10^{10}$	2670	0
5	$CH_3OH + OH$	$—CH_2OH + H_2O$	$3,0 \cdot 10^{11}$	3000	0
6	$CH_2OH + O_2$	$—CH_2O + HO_2$	$5,0 \cdot 10^7$	0	0
7	$CH_2OH + M$	$—CH_2O + H + M$	$2,5 \cdot 10^{11}$	14600	0
8	$CH_4 + O_2$	$—CH_3 + HO_2$	$8,0 \cdot 10^{11}$	28300	
9	$CH_3 + O_2$	$—OH + CH_2O$	$2,0 \cdot 10^7$		
10	$CH_4 + OH$	$—CH_3 + H_2O$	$6,0 \cdot 10^{11}$	6290	0
11	$HCO + O_2$	$—CO + HO_2$	$1,0 \cdot 10^{11}$	3434	
12	$CH_4 + HO_2$	$—CH_3 + H_2O_2$	$2,0 \cdot 10^{10}$	9091	
13	$CH_4 + H$	$—CH_3 + H_2$	$2,2 \cdot 10^1$	4400	3
14	$CH_4 + O$	$—CH_3 + OH$	$2,1 \cdot 10^{10}$	4560	0
15	$CO + HO_2$	$—CO_2 + OH$	$1,0 \cdot 10^{14}$	11616	0
16	$CH_2O + O_2$	$—HCO + HO_2$	$1,0 \cdot 10^{11}$	16162	
17	$HCO + M$	$—CO + H + M$	$5,0 \cdot 10^{11}$	9570	
18	$CO + OH$	$—CO_2 + H$	$4,0 \cdot 10^9$	4030	
19	$CH_2O + OH$	$—H_2O + HCO$	$5,4 \cdot 10^{11}$	3170	0
20	$CH_3 + O$	$—CH_2O + H$	$1,0 \cdot 10^{11}$		
21	$CH_2O + H$	$—HCO + H_2$	$1,35 \cdot 10^{10}$	1890	
22	$CH_2O + O$	$—HCO + OH$	$5,0 \cdot 10^{10}$	2300	
23	$HCO + OH$	$—CO + H_2O$	$1,0 \cdot 10^{11}$	0	0
24	$CH_3 + O_2$	$—H_2 + CO + OH$	$4,0 \cdot 10^9$	9091	0
25	$CO + O + M$	$—CO_2 + M$	$6,0 \cdot 10^7$	0	0
26	$CH_2O + M$	$—CHO + H + M$	$1,0 \cdot 10^{11}$	18500	
27	$CHO + H$	$—CO + H_2$	$2,0 \cdot 10^{11}$	0	0
28	$CHO + O$	$—CO + OH$	$1,0 \cdot 10^{11}$	0	0
29	$H + H + M$	$—H_2 + M$	$1,0 \cdot 10^{12}$	0	-1
30	$O + O + M$	$—O_2M$	$1,0 \cdot 10^8$	0	0
31	$O + H + M$	$—OH + M$	$3,0 \cdot 10^8$	0	0
32	$H + O_2$	$—OH + O$	$2,2 \cdot 10^{11}$	8462	0
33	$O + H_2$	$—OH + H$	$1,8 \cdot 10^7$	4482	1
34	$OH + H_2$	$—H_2O + H$	$2,2 \cdot 10^8$	2593	0
35	$H + OH + M$	$—H_2O + M$	$1,5 \cdot 10^{11}$	0	$-0,5$
36	$2OH$	$—H_2O + O$	$6,3 \cdot 10^9$	553	0
37	$H_2 + O_2$	$—2OH$	$1,36 \cdot 10^{10}$	24318	0
38	$H + HO_2$	$—2OH$	$7,3 \cdot 10^{11}$	0	0
39	$N_2 + M$	$—2N + M$	$2,0 \cdot 10^{18}$	113316	$-1,5$
40	$NO + M$	$—N + O + M$	$5,5 \cdot 10^{17}$	75544	$-1,5$
41	$NO + O$	$—N + O_2$	$1,55 \cdot 10^6$	19439	1
42	$O + N_2$	$—NO + N$	$1,30 \cdot 10^{11}$	37974	0
43	$N_2O + M$	$—N_2 + O + M$	$1,0 \cdot 10^{12}$	30722	0
44	$2NO$	$—N_2O + O$	$2,6 \cdot 10^9$	32130	0
45	$NO + O_2$	$—NO_2 + O$	$7,8 \cdot 10^8$	22930	0
46	$N_2 + O_2$	$—NO + NO$	$9,1 \cdot 10^{20}$	64970	$-2,5$
47	$NO_2 + M$	$—O + NO + M$	$6,0 \cdot 10^{18}$	36060	$-1,5$
48	$NO + O_3$	$—NO_2 + O_2$	$8,9 \cdot 10^8$	1330	0
49	$HNO + H$	$—NO + O_2$	$4,5 \cdot 10^9$	0	0
50	$N_2O + H$	$—N_2 + OH$	$3,0 \cdot 10^{11}$	8080	0
51	$HNO + OH$	$—NO + H_2O$	$3,0 \cdot 10^9$	1200	0,5
52	$H + NO + M$	$—HNO + M$	$5,4 \cdot 10^9$	-300	0
53	$HNO + NO$	$—N_2O + OH$	$6,14 \cdot 10^9$	17222	0
54	$NH_3 + NO$	$—NH_2HNO$	$1,0 \cdot 10^7$	0	0
55	$NH_2 + H + M$	$—NH_3 + M$	$4,8 \cdot 10^8$	-8300	0
56	$NH_3 + H$	$—NH_2 + H_2$	$5,0 \cdot 10^8$	1000	0,5
57	$N + OH$	$—H + NO$	$1,2 \cdot 10^{10}$	0	0
58	$NH_3 + N$	$—NH_2 + NH$	$5,0 \cdot 10^8$	1010	0,5
59	$NH_3 + OH$	$—NH_2 + H_2O$	$5,4 \cdot 10^{12}$		0
60	$NH_3 + O$	$—NH_2 + OH$	$1,0 \cdot 10^9$	2475	0
61	$NH + HNO$	$—NH_2 + NO$	$2,0 \cdot 10^8$	1010	0,5

(**Fig. 2.1** Contd.)

No.	Reaction		$A[m^{3k}/k\,mol^k \cdot s]$	E/R_m	n
62	$HNO_3 + M$	$-OH + NO_2 + M$	$1,6 \cdot 10^{12}$	15450	0
63	$NH_2 + OH$	$-NH + H_2O$	$3,0 \cdot 10^7$	656	0,679
64	$NH_2 + NO$	$-N_2 + H_2O$	$1,0 \cdot 10^{10}$	0	0

k = 1 for bimolecular reactions
k = 2 for trimolecular reactions

Fig. 2.1 Reaction mechanism of the methanol-air reaction [2.1]

2.2 Toxicity and effects of pollutants on mankind

Emissions represent the pollutants emitted by the engine and are indicated e.g. in grams per hours or grams per kWh. Imissions, on the other hand, result from emissions and allow statements to be made on the distribution of specific substances in the environment. As another descriptor, concentration is used to indicate the share per weight unit, e.g. in % by volume or ppm, of a particular substance in the exhaust. The maximum workplace concentration (MAK value) of a substance is indicated e.g. in ppm or mg/m^3.

- *Carbon monoxide:* A colorless and odorless gas. Its adherence to haemoglobine (the O_2 carrier in the blood) is far stronger (factor 240) than that of oxygen. Even low CO concentrations may therefore be sufficient to cause suffocation. The MAK value is $33\,mg/m^3$ [2.36].
- *Unburned hydrocarbons:* Depending on their composition, they have a more or less narcotic effect and irritate man's mucous membranes. Certain components have a carcinogenic effect (aromates, e.g. 3,4 benzapyrene, benzene). The effect of anoxidised hydrocarbons (aldehydes) on aromates is comparable to that of unburned hydrocarbons.
- *Nitrogen dioxide:* A gas with a sharp smell and red-brown color. Low concentrations are sufficient to cause lung irritation, tissue damage and irritation of mucous membranes. A risk of acid formation is present in the same way as with nitrogen monoxide. The MAK value is $9\,mg/m^3$ [2.36].
- *Nitrogen monoxide:* An odorless gas that modifies the function of the lungs. It irritates man's mucous membranes and oxidises with O_2 to produce NO_2. Risk of nitric acid formation. As nitrogen monoxide is instable under ambient conditions and will change into nitrogen dioxide, the MAK value also is $9\,mg/m^3$ [2.45].
- *Aldehydes:* Components with a sharp smell and narcotic effect. Some of these compounds are considered to cause cancer. The MAK value, e.g. of formaldehyde, is $0.6\,mg/m^3$ [2.36].
- *Lead:* A cell poison that reduces O_2 absorption of the blood. The MAK value is $0.1\,mg/m^3$ [2.45].
- *Sulphur dioxide:* An odorless gas with a sharp smell, causing irritation of mucous membranes. Produces sulphuric acid under the action with water (highly caustic effect): The MAK value is $2\,ml/m^3$ [2.45].
- *Particulates:* Diesel engines are the target of increased criticism since they generate particulate emissions (carcinogenic potential) and sulphur dioxide emissions that contribute to environmental damage known as "tree death". Part of the particulates can enter the lungs and are dangerous since they deposit substances that constitute a health hazard. Particulates also contain soot that may be present either as pure carbon or with deposited hydrocarbons. Recent findings indicate that the soot core may have a carcinogenic effect.

  Other investigations [2.37] [2.38] [2.39], however, underline the theory that this carcinogenic potential is due not to diesel soot cores but, in a more general context, to

2.3 Causes of pollutants in engines

Type of risk	Deaths	Death risk per million inhab.
Motor traffic	45901	194,37
Fall	12001	50,82
Fire	4938	20,91
Drowning	4407	18,66
Drugs, medicine	3612	15,29
Swallowing	1663	7,04
Guns	1649	6,98
Air traffic	1428	6,05
Implements	1288	5,45
Dropping objects	903	3,82
Electrical shock	802	3,40
Alcohol poisoning	305	1,29
Lightning	85	0,34
Poisonous bite / insect bite	49	0,21
Dog bite	15	0,06
Fireworks	11	0,05
Smokers, estimated		800
Non-smokers, estimated		70
Diesel fumes, estimated		
City inhabitants, general		0,5
Living next to urban highway		6
Road workers		35

- Time, March 23, 1989
- Inhabitants: 236 million, Britannica BY 1985
  Risk = Death divided by no. of inhabitants
- Cuddihy et al 1984 acc to Meclellan 1985
  Risk referred to affected group

Fig. 2.2 General death risks in the USA [2.12]

particulates in the breathing air that are not necessarily specific to diesel exhaust fumes. Epidemiologic studies have so far failed to prove the carcinogenic effect of diesel exhaust on man [2.37].

The risk of falling mortally sick due to diesel engine exhaust seems to be extremely small, as indicated by a study carried out in the USA. It is about as small as being killed by lightning (*Fig. 2.2*) [2.12]. Yet further measures are required to reduce particulate emissions.

- *Carbon dioxide:* Carbon dioxide has no direct effect on man at the concentrations present in engine operation but contributes to long-term environmental damage caused by atmospheric changes (greenhouse effect). Excessive concentrations may lead to suffocating. The MAK value of carbon dioxide is 9,000 mg/m^3 [2.36].

Further information on toxicity, maximum and minimum workplace concentrations (MAK and MIK values) is found in the referenced literature, e.g. [2.40].

2.3 Causes of pollutants in engines

2.3.1 Causes of pollutants in gasoline engines

The various pollutant components (except nitrogen oxides) are primarily determined by the type of fuel and its additives as well as by engine oils burned in the combustion chamber. The level of emissions is influenced primarily by engine behavior.

The following section deals with raw emissions, i.e. emissions that do not undergo exhaust aftertreatment.

A typical exhaust composition is shown in *Fig. 2.3*. This is the result of an analysis of the exhaust gas produced during a complete ECE test cycle (cf. Chapter 8) using the reference fuel specified for this purpose.

Apart from CO_2 that may also be considered a pollutant, approx. 1% of the total exhaust mass emitted in the ECE test cycle has to be considered as pollutants (reduced by one order of magnitude after catalytic converter treatment).

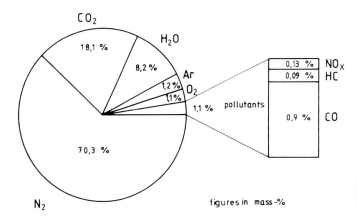

Fig. 2.3 Mean exhaust composition without catalytic converter in ECE test (medium-size vehicle) [2.33]

In additon, the effect of the major engine parameters on pollutant emissions is highlighted. The essential parameter is the air/fuel ratio λ. This unit is defined as

$$\lambda = m_L/(M_{fuel} \cdot m_{Lstoich}).$$

The unit $m_{Lstoich}$ represents the stoichiometric air mass. This quantity depends on the type of fuel and its composition and is approx. 14.6 kg of air per one kg of fuel in the case of gasoline and diesel fuel.

Along with the specific fuel consumption, the basic dependence of pollutant concentrations on the air/fuel ratio is indicated in *Fig. 2.4* for the essential pollutant components produced in a spark-ignition engine.

As is widely known, a conventionally operated spark-ignition engine operates on the principle of quantity management, i.e. the air/fuel ratio my be varied only within very narrow limits. This is due to the relatively narrow misfire limits of the air-fuel mixture. In engine operation, the air/fuel ratio is approx. between $\lambda = 0.85$ and $\lambda = 1.2$ if no lambda control is present. Load management is accomplished by varying the mixture mass supplied to the cylinder using the throttle.

Further reasons governing the process of formation of major pollutant components as a function of the air/fuel ratio are highlighted in the following section.

2.3.1.1 Pollutant components

- *Carbon monoxide:*
  This component can be influenced primarily by the air/fuel ratio λ. A lack of air ($\lambda < 1$) results in high CO concentrations as not enough oxygen is available to oxidise all C molecules sufficiently into CO_2. As excess air increases, the CO concentrations decrease.

  Carbon monoxide emissions can be only influenced to an insignificant extent by engine-related parameters such as ignition timing, compression ratio, engine speed and injection timing. This is explained by the fact that the CO recombination reactions that occur during the expansion phase are primarily dependent on pressure [2.2]. The pressure during the expansion phase, however, is more or less independent of the above parameters. CO emissions therefore are not dependent on the above parameters to any significant extent [2.2]. One reaction that is of relevance to CO formation is described by the "water gas equation":

$$CO + H_2O \Leftrightarrow CO_2 + H_2.$$

2.3 Causes of pollutants in engines

This gross reaction is a sufficiently accurate description of the entire CO production process in the oxygen deficiency range. In the excess air range, CO concentrations are relatively low and decrease further as the air/fuel ratio increases. The CO present in this range may be traced to a local lack of homogeneity of the air-fuel mixture and to reaction processes that occur near the wall or to freezing of rections as an increasing amount of excess air becomes available. The below gross conversion equation is of major importance to CO formation in the excess air range:

$$CO + \tfrac{1}{2}O_2 \Leftrightarrow CO_2$$

- *Unburned hydrocarbons:*
If a hydrocarbon molecule is combusted under "ideal" conditions, only a relatively small quantity of unburned hydrocarbons (referred to hereafter as HC) is released. Unburned hydrocarbons will remain only in those areas where the flame does not popagate. This may be the case in a number of areas in the combustion chamber of a spark-ignition engine, such as gaps in the combustion chamber in the vicinity of the cylinder head gasket, the piston top land, piston rings and the spark plug areas as well as in the case of poorly shaped squish areas etc. HC emissions are also formed if the heat exchange from the gas in the vicinity of the wall is large enough to extinguish the flame (quench effect) [2.42]. Another effect that may prevent the hydrocarbons from being combusted is caused by approaching the misfire limits in the lean or rich range of the mixture. If air or fuel mass variations cause the mixture to become sufficiently lean to prevent it from being ignited, significant hydrocarbon emissions are produced.

Since this range usually is not relevant thanks to state-of-the-art engine management used in engines operated with lambda control, HC emissions may in most cases be traced to quenching of the flame in combustion chamber gaps and at the cold cylinder walls.

Detachment of the lubricating film is another cause for the presence of unburned hydrocarbons in the exhaust gas. The wall film consisting of unburned fuel and lubricating oil hydrocarbons is detached and is evacuated into the exhaust system during the charge changing process [2.19]. A certain amount of the hydrocarbons released in this process is combusted during the post-oxidation phase in the exhaust section. This afterburning process requires a certain temperature level and an appropriate oxygen content. The process of HC emissions formation is shown schematically in *Fig. 2.5.*

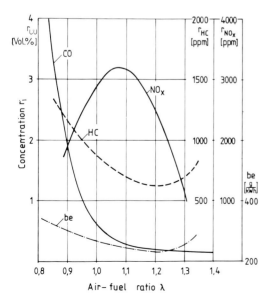

Fig. 2.4 Pollutant concentrations and specific fuel consumption as a function of air/fuel ratio

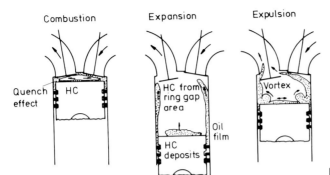

Fig. 2.5 Formation of HC emissions [2.19]

The composition of unburned hydrocarbons may vary widely. Essential components are e.g. aromatic substances (benzene, toluene, ethyl benzene) as well as olefins (e.g. propene, ethylene) and paraffins (e.g. methane).

- *Nitrogen oxides:*
The diagram of NO_x concentrations vs air/fuel ratio (*Fig. 2.3*) shows that the nitrogen oxides reach a maximum at a slight excess of air. Generation of these components is enhanced by high, local peak temperatures and a corresponding excess of air. Such conditions are present at $\lambda = 1.05$ to 1.1. The high temperatures encourage splitting of N_2 and O_2 into their atomic constituents. Excess air ensures that sufficient oxygen is present.

All engine-related parameters having an effect on the above boundary conditions will also influence NO_x emissions (e.g. load, λ, ignition angle, compression ratio). NO emissions account for approx. 90 to 98% of all NO_x emissions during engine operation. The reaction mechanism relevant for NO_x formation is governed primarily by the Zeldovic mechanism. In addition, the below reactions are of importance [2.7]:

$$O_2 \Leftrightarrow 2O$$
$$OH + N \Leftrightarrow NO + H$$

Reactions that generate nitrogen oxide occur relatively slowly. NO is formed in significant quantities only behind the flame front in the combusted matter [2.2].

- *Aldehydes:*
Aldehydes are hydrocarbons with additional embedded oxygen atoms. These O–H–C compounds are produced mainly during the combustion of fuels with high oxygen contents, e.g. alcohols. Some aldehydes are highly odorous. One major representative of this pollutant group is formaldehyde (HCHO) which is already subject to regulations in California.

- *Lead compounds:*
Lead emissions of spark-ignition engines are caused exclusively by lead additives contained in gasoline. Lead is usually found in anti-knock additives based on chlorine and bromium compounds that are used to reduce the high boiling temperature of lead. The use of lead additives is decreasing rapidly as they "contaminate" the catalytic converters used today (cf. Chapter 5).

2.3.2 Causes of pollutants in diesel engines

Following the introduction of three-way catalytic converters with closed-loop control for spark-ignition engines, the toxic exhaust emissions were reduced by one order of magnitude over non-emission control versions, and environmental concern rapidly shifted to focus on the emission characteristics of diesel engines. One major reason was that no equivalent measures were available to reduce nitrogen oxide emissions of diesel engines.

State-of-the-art engine development and exhaust aftertreatment systems as well as fuel modifications, however, open up new perspectives for reducing pollutant emissions of diesel engines drastically (cf. Chapter 3, 4, and 5).

2.3.2.1 Pollutant components

Load management of diesel engines is accomplished by modifying the fuel quantity supplied, whereas the air quantity per work cycle is drawn in without any throttling. The air to fuel mixture ratio is therefore modified to achieve load management.

This means that the mixture is made leaner or richer, respectively. Depending on friction power and the combustion process of the engine, the mixture setting is made leaner up to levels of $\lambda = 10$ (see *Fig. 2.6*). This diesel-engine processes is referred to as quality management.

In the same manner as with spark-ignition engines, the air/fuel ratio of the diesel engine has a significant impact on the level of pollutant concentrations but this parameter is not freely available for minimizing pollution. *Figure 2.7* shows an example of the dependence of major pollutant concentrations on the air/fuel ratio in the case of a direct-injection diesel engine.

- *Carbon monoxide:*
  The mean air/fuel ratio present in the combustion chamber per cycle is far higher in the diesel engine than in the SI engine. Due to a lack of homogeneity of the mixture built up by stratification, however, extremely "rich" local zones are present. This produces high CO concentrations that are reduced to a greater or lesser extent by post-oxidation.

  When the λ coefficient increases, i.e. when the excess-air ratio increases, dropping temperatures cause the post-oxidation rate to be reduced (*Fig. 2.7*). The reactions "freeze up". The CO concentrations of diesel engines therefore are far lower than in SI engines. The basic principles of CO formation, however, are the same.

- *Unburned hydrocarbons:*
  Since the air-fuel mixture is not homogeneous throughout, extremely high excess air ratios are present in certain zones during the diesel combustion process. The higher the air/fuel ratio, the lower is the local temperature. This means that chemical reactions

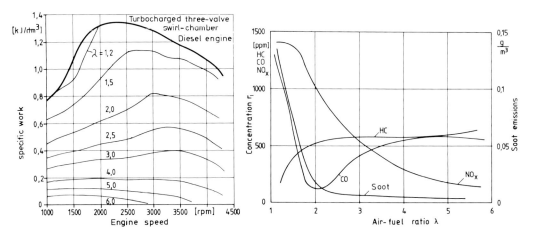

Fig. 2.6 λ map of a diesel engine

Fig. 2.7 Exhaust concentrations as a function of air/fuel ratio [2.19]

proceed fairly slowly or may even "freeze up", thus leading to increased HC emissions. On the whole, HC concentrations of diesel engines are lower than those of spark-ignition engines (see *Figs. 2.3* and *2.7*).

- *Nitrogen oxides*:
  NO_x concentrations are lower than in spark-ignition engines; the share of NO_2 in the NO_x emissions is slightly higher. When comparing an SI engine with three-way closed-loop catalytic converter with a diesel engine with oxidation catalyst, it is found that NO_x concentrations of the SI engine are lower.

The type of combustion process has a significant effect on nitrogen oxide formation.

Diesel engine with divided combustion chamber: In engines with a divided combustion chamber, combustion initially occurs in the precombustion or swirl chamber under conditions of extreme oxygen deficiency. This generates high temperatures, yet NO_x levels are low due to a lack of air and, hence, oxygen. This process is reversed in the main combustion chamber. Extreme excess air ratios and, hence, low temperatures also result in low NO_x formation.

Diesel engine with direct injection: The direct-injection diesel engine does not have the above features that keep NO_x emissions low. As a result, NO_x formation is approximately twice as high as with an engine with divided combustion chamber. When exhaust recirculation (EGR) systems are used, both systems show virtually identical NO_x levels as the direct-injection diesel engine is far more compatible with large recirculated exhaust quantities (more than 50% of the fresh charge).

- *Sulphur compounds*:
  Sulphur compounds are caused exclusively by the sulphur content of the fuel. According to German regulations, this content may be as much as 0.2 weight percent, although it currently is below 0.2 weight percent and will have to be reduced to below 0.05 weight percent by October, 1996, throughout Europe. When combined with the water produced during the combustion process, SO_2 produces sulphuric acid. Sulphur compounds cause problems with regard to acid rain and particulate formation via sulphates.

- *Particulates*:
  According to the California Air Resources Board (CARB), particulate matter is defined as follows [2.9]:

  Particulates are all exhaust components (with the exception of condensed water) that are deposited on a defined filter after having been diluted with air to a temperature below 51.7°C.

  Basically, soot emissions also are part of particulate emissions. Soot formation [2.43] occurs at extreme air deficiency. This air or oxygen deficiency is present locally inside diesel engines. It increases as the air/fuel ratio approaches a value $\lambda = 0$. Soot is produced by thermal cracking of long-chain molecules at oxygen deficiency. A separation of hydrogen leads to C-structures showing an increasing lack of hydrogen. Acetylene and other polymerization processes lead to formation of molecules rich in carbon that form soot particulates [2.11] [2.12] [2.44]. Once soot has formed, it can be oxidised only to a limited extent. The process described here is shown schematically in *Fig. 2.8*. The logarithm of the molecular mass is shown as a function of the hydrogen contents of the hydrocarbon components. Soot formation produces molecules with an increasingly low hydrogen content and higher weight that will finally agglomerate to form soot particulates.

  Particulates consist of solid (organically insoluble) and liquid (organically soluble) phases.

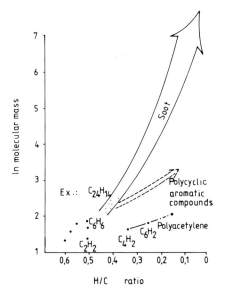

Fig. 2.8 Potential soot formation process

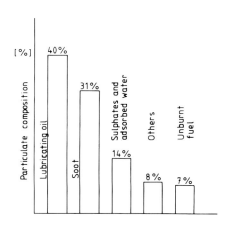

Fig. 2.9 Particulate composition of diesel engine exhaust

The solid phase consists of

- soot in the form of amorphous carbon, ash, oil additives, corrosion products and abrasion products.
- sulphates and its molecule-bound water.

The liquid phase consists of

- fuel and lubricant contents that are, in most cases, combined with soot. The hydrocarbons contained in the hot exhaust still are largely gaseous and are converted into a liquid, organically soluble phase (particulates) only after having been cooled by turbulent intermixing with air.

The size of such particulates is approx. 0.01 to 1 µm and above. Most particulates have a size below 0.3 µm and some of them can therefore penetrate into the lungs. Particulate composition is largely dependent on the operating point and the combustion process. *Figure 2.9* shows a typical composition of diesel engine exhaust gas.

2.4 Causes and effects of pollutants in the atmosphere

Climate is the totality of all meteorological conditions over a longer period that are generated due to interaction within the overall system. The climate mechanism is made more complicated due to the fact that the individual subsystems, i.e. hydrosphere, lithosphere, biosphere and atmosphere are not related in a linear manner but act in an extremely non-linear manner, cf. *Fig. 2.10* [2.20] [2.21] [2.22] [2.23].

Feedback may intensify or attenuate the worldwide processes. The different reaction times within the existing relationships also tend to complicate observations and forecasts. External impacts superimpose internal effects of the subsystems and include, for example, changes due to sun radiation and volcanic eruptions as well as effects caused by human activity on the nature of the earth surface, the hydrologic cycle and cycles of atmospheric trace gases.

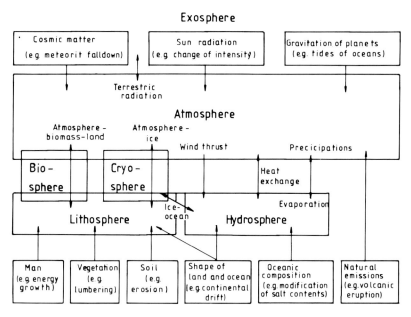

Fig. 2.10 Interrelationship of subsystems of the climatic mechanism [2.20]

When categorizing the effects of pollutants in the biosphere, a distinction should be made among the following phenomenons:

- Smog,
- Acid rain,
- Ozone depletion ("ozone hole"),
- Ozone pollution,
- "Greenhouse effects".

Some of these are of a global nature, such as the "greenhouse effect" and depletion of the ozone layer. Ozone pollution near the ground level, smog and "acid rain" occur mostly near emittent sources or in areas with dense population, although they may affect large areas.

One common feature of all effects is that they are caused largely by anthropogenic emissions (industrial plants, domestic firing, traffic). This kind of anthropogenic emissions of trace gases and aerosols has increased virtually ever since the beginning of this century along with the increasing level of industrialization [2.32]. The emitted pollutants enter the atmosphere and are modified there by physical and chemical conversion processes. When discussing the effects of pollutants, including not only those emitted by internal-combustion engines, in terms of changes of the air quality and the resulting implications, several topics are affected. These topics essentially include effects on ozone depletion in the stratosphere ("ozone hole") [2.32], formation of photooxidants in the troposphere, the "greenhouse effect" and "acid rain". *Fig. 2.11* shows the individual layers of the earth atmosphere with reference to the altitude and the temperatures of the individual layers.

The above processes should not be treated in an isolated fashion. They rather involve a variety of chemical and physical processes that affect each other and sometimes are of highly complex nature.

Of significant importance worldwide are conversion processes that are based on photochemical reactions and cause what is referred to as photochemical smog as well as those processes that contribute to the greenhouse effect (cf. Chapter 7). In addition to automotive exhaust emissions, a number of other emission sources, e.g. industry, domestic

2.4 Causes and effects of pollutants in the atmosphere

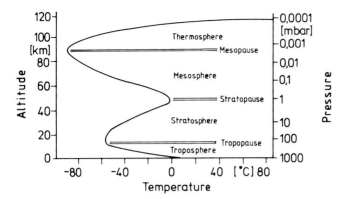

Fig. 2.11 Layers of the earth atmosphere [2.22]

firing and energy generation, are responsible for the ozone depletion in the stratosphere (chlorofluorohydrocarbons = CFC) as well as for acid rain (SO_2) and the greenhouse effect (CO_2).

While processes causing acid rain are largely based on reactions involving sulphur dioxide and nitrogen oxides, photochemical smog involves, among others, the NO_x, CO and HC components. In this process, sun radiation produces substances known as photooxidants. Particular attention should be paid to ozone (O_3) in this respect. This is essentially dictated by three reasons:

- Ozone is the most important component in terms of its percentage share.
- Ozone photolysis starts formation of radicals that enable the reactions to continue.
- Ozone has a major impact on the climate and contributes significantly to the greenhouse effect in the atmosphere.

The radicals present, especially the hydroxyl radical, are capable of modifying further components.

The ozone production processes occur in a different manner in the troposphere and stratosphere, respectively. As opposed to other trace gases affecting the climate, atmospheric ozone has only a short retention time in the atmosphere. Another important fact is that its concentration increases in the troposphere, especially in the Northern hemisphere, whereas it decreases in the stratosphere worldwide (ozone hole). Also refer to *Fig. 2.12* [2.15]. This figure shows the percent ozone modification in the stratosphere and in the troposphere until 2050. The data were obtained from model calculations that are based on an annual 4% increase of CFC emissions.

Emissions caused by fossil energy carriers, e.g. emissions of passenger cars, have not been taken into account. It may therefore be assumed that ozone emissions in the lower troposphere will increase locally beyond the figures used in this model calculation.

Ozone has a toxic effect in the biosphere, and it is assumed that, when combined with other photooxidants, it causes new types of forest damage [2.13].

Ozone also affects man, i.e. ozone that is present in sufficient concentrations may cause irritation of eyes and mucous membranes as well as breathing difficulty under heavy bodily effort. According to the German Federal Health Ministy, the maximum ozone concentration at which adverse effects on health may be ruled out positively is $120 \, g/m^3$ [2.14]. Peak values of 130 to $150 \, g/m^3$ and above occur relatively frequently in the summer months in European latitudes [2.14].

In other areas, especially in Los Angeles, far higher concentrations have been measured. *Figure 2.12* shows hourly mean values between 600 and $700 \, g/m^3$ [2.17].

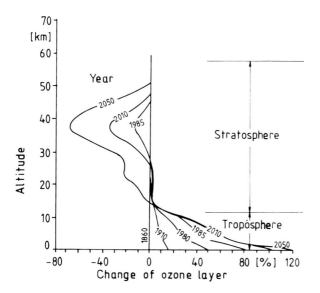

Fig. 2.12 Results of a model calculation on the modifications of the ozone profile based on an annual 4% increase of CFC emissions [2.15]

The presence of ozone in the troposphere is due to the fact that it is carried in from the stratosphere and is formed in the troposphere. Nitrogen oxides NO and NO_2 as well as components such as CO, CH_4 and higher hydrocarbons are of relevance to ozone formation. The ratio of NO_x to O_3 is of particular importance.

Ozone is produced by the reaction of molecular oxygen with atomic oxygen. Atomic oxygen is produced e.g. by NO_2 photolysis

$$NO_2 + h\cdot v \Leftrightarrow NO + O.$$

Decomposition is most efficient in the following reaction:

$$NO + O_3 \Leftrightarrow NO_2 + O_2.$$

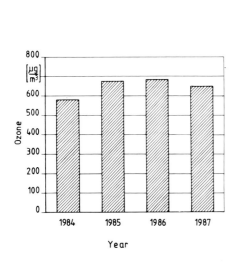

Fig. 2.13 Hourly mean ozone value in Los Angeles [2.17]

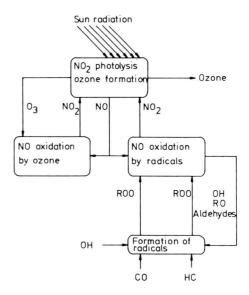

Fig. 2.14 Photochemical process of ozone formation [2.17]

The above reactions, however, are not the only ones that have an effect on ozone formation. An essential contributing factor is the hydroxyl radical (OH) that is formed, for example, when water steam reacts with atomic oxygen. The entire process of photochemical processes is shown schematically in *Fig. 2.14* in accordance with [2.17].

Some components (e.g. SO_2 and NO_x) cause the pH value of precipitations to change during tropospheric conversion. This phenomenon known as "acid rain" is known to be part of a system that affects vegetation adversely. Acidic matter in the lower layers of the atmosphere also is an important contributing factor to what is known as "London smog", a phenomenon, however, that has ceased to occur after the changeover was made from coal heating to natural gas heating.

A characteristic feature of this type of smog was severe pollution of the air occurring during inversion weather conditions (dust, smoke, particulates, chemically active substances).

The gases released into the atmosphere may be subdivided into substances that affect the climate either directly or indirectly and are both generated during engine combustion processes. The gases affecting the climate directly that are not generated exclusively by engine combustion processes are of varying importance and include CO_2, CH_4, N_2O, halogenated hydrocarbons and ozone. Motor vehicle traffic accounts for approx. 20% of all CO_2 emissions. Products affecting the climate indirectly that are generated by the internal-combustion engine are essentially CO, NO, NO_2 and volatile organic compounds.

The most important carbon monoide decomposition process in the troposphere involves the reaction with hydroxyl (OH). To a lesser extent, CO is also carried into the stratosphere where it is oxidised by photochemical reactions. The anthropogenic volatile hydrocarbon compounds (see *Fig. 2.15* for an overview of the major representatives and residence times of these compounds [2.13]) are produced mostly by combustion of fossil fuels and during the transport of such substances as well as during the combustion of biomass, respectively.

Investigations [2.18] have shown that the presence of both components, i.e. non-methane hydrocarbons and NO_x, is important to allow ozone formation to occur in layers close to the earth. If, for example, only NO_x emissions were reduced, ozone formation

Gas	Residence time in days	
	Tropical region	global
Ethane	20,00	61,0
Methanol	6,50	15,0
Acetylene	5,50	13,0
Benzene	4,50	14,0
Propane	3,60	11,0
n-butane	2,10	7,5
i-butane	2,20	8,0
Dimethyl ether	1,90	7,0
Ethyl alcohol	1,80	5,9
n-pentane	1,50	4,8
Toluene	0,90	2,8
Ethyl benzene	0,75	2,5
Ethylene	0,70	1,9
Formaldehyde	0,50	0,9
o-xylene	0,35	0,7
Acetaldehyde	0,30	0,6
Propylene	0,20	0,4

Fig. 2.15 Tropospheric residence times or organic compounds [2.13]

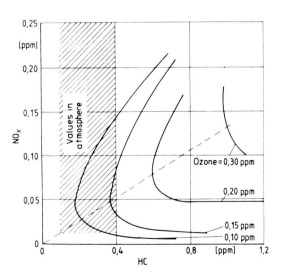

Fig. 2.16 Influence of HC and NO_x on ozone formation [2.18]

might increase in certain cases (cf. *Fig. 2.16*). This is why it is important to reduce both components, i.e. HC and NO_x, in metropolitan areas.

In order to reduce the gases affecting climatic conditions yet further, both a drastic reduction of fuel consumption and a global introduction of efficient emission control systems will be required, e.g. improved three-way catalytic converters for spark-ignition engines and diesel catalytic converters (oxydation or lean-burn catalysts and, possibly, soot traps) for diesel engines.

2.4.1 CO_2 and the climate

One effect that has gained particular importance due to CO_2 is what is referred to as the greenhouse effect. Human activities modify the gas composition of the atmosphere. This disturbance of the natural radiation exchange results in changes to the climate [2.24] [2.25] [2.26] [2.27] [2.28] [2.41].

The energy supplied to the earth from outside essentially consists of short-wave radiation. The pollutants present in the atmosphere, especially CO_2, but also water vapor, O_3, NO_2, aerosols and trace gases, allow this radiation to pass virtually undisturbed. Upon reaching the earth surface, short-wave radiation is transformed into heat and is emitted again as long-wave heat radiation. This infrared radiation is absorbed and reflected, respectively, by the pollutants. The result of this process, however, is that the temperature increases in the atmosphere and troposphere, respctively.

Both combustion processes and the relevant agricultural activities contribute to modifying the trace gas content in the atmosphere. Combustion gases play a major role in the resultant warming of the troposphere. Increasing emissions of carbon dioxide, chlorofluorocarbons, methane, ozone and nitrous oxide intensify the natural greenhouse effect and, hence, contribute to heating up the air layer near the ground level. *Figure 2.17* shows the percent contributions of the individual emissions to the greenhouse effect and the percent share of carbon dioxide emissions identified by geographic regions [2.21].

Figure 2.17 indicates that CO_2 is the most important contributing factor to an increased greenhouse effect. Efforts to reduce this component are therefore of major importance. *Figure 2.18* shows the extent to which the carbon dioxide content in the atmosphere has changed.

Based on measurements on Mount Mauna Loa, Hawaii, an increase of 315 ppm in 1958 to 350 ppm in 1988 was observed [2.29]. The mean value computed for recent decades shows that CO_2 concentrations have increased by approx. 0.5% per year. Various scenarios indicate varying forecasts for future increases and the resulting temperature increase.

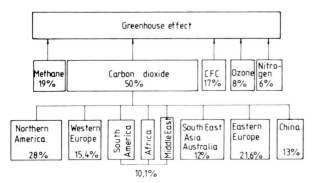

Fig. 2.17 Contributing factors to the greenhouse effect [2.21]

2.4 Causes and effects of pollutants in the atmosphere

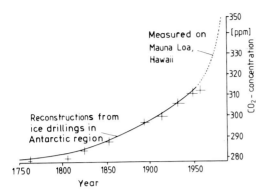

Fig. 2.18 Increase of CO_2 concentrations in the atmosphere [2.19]

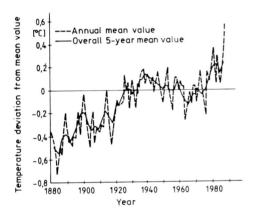

Fig. 2.19 Deviation from average temperature [2.30]

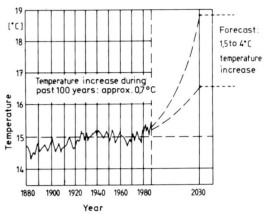

Fig. 2.20 Forecast of temperature increase [2.30]

Investigations of the ground level global average temperatures of the past century reveal a tendency towards increasing temperatures. *Figure 2.19* shows the temperature deviations from the mean value of the ground level global average temperature (15 °C) from 1880 to 1985. This shows a worldwide temperature increase of 0.7 °C over the past 100 years [2.30].

Due to the ever-increasing worldwide energy requirements, the annual increase of the carbon dioxide content in the atmosphere will continue along with the increasing energy requirements. Pessimistic forecasts that take other gases affecting the greenhouse effect into account indicate that the global temperature will increase by approx. 1.5 °C to 4 °C over the next 50 years (*Fig. 2.20*) [2.31].

3 Design features which influence pollutant emissions and fuel consumption in four-stroke engines

The efficiency of an internal-combustion engine and, hence, its specific fuel consumption are largely dependent on the process characteristics of each cycle. As the internal-combustion process in the engine cannot really be described accurately by computation, comparison processes and descriptions of part systems are required. These comparison processes allow basic statements to be made e.g. on the effect that the air/fuel ratio or the compression ratio has on efficiency. Although the chemical transformation of the process matter during the combustion process is taken into account in this type of process control (e.g. ideal-engine process), the statements on pollutant behavior and/or fuel consumption will only reveal basic trends. This is due to the required simplification of boundary conditions of such processes, e.g. parameters such as isentropic compression and expansion, combustion products in chemical equilibrium and heat-sealed walls. In many cases, these conditions are far removed from the real-life processes occurring in an engine. Comparison processes are therefore not dealt with further in this study.

Pollutant emissions and fuel consumption of both spark-ignition and diesel engines may be reduced in a number of ways [3.9] [3.60]. Three basic approaches are commonly used:

- Design parameters, e.g. cylinder head and combustion chamber design, including spark plug or glow plug position, design of injection system, injector position, number of valves, displacement (swept volume), bore-to-stroke ratio, compression ratio, inlet and exhaust port shape etc.
- Operational parameters. They include e.g. engine management, mixture formation and mixture control, ignition timing, injection duration and injection timing, valve timing, internal exhaust recirculation based on modified engine timing, external exhaust recirculation, introduction of a latent heat accumulator to utilize engine "waste heat" [3.39], fuel supply cutoff in overrunning mode etc.
- Exhaust gas aftertreatment using catalytic converter systems (including heated catalytic converters), secondary air injection, insulation of exhaust manifold and exhaust system, thermal reactors, filters (traps) or particulate retention systems etc.

Emission characteristics are influenced heavily by transient operating conditions and the operating period at which the engine has not yet reached its operating temperature. When running the stipulated test cycles (cf. Chapter 8), the intial 60 to 80 seconds have a decisive effect on whether the emission standards will be complied with. Cold intake pipes and combustion chamber walls, the higher frictional power to be overcome and a catalytic converter that has not yet reached its operating temperature are some of the factors that contribute to sharply increased emissions, particularly of HC and CO. This is where particular demands are placed on mixture formation and exhaust aftertreatment systems. At the same time, mixture formation under transient operating conditions is subject to additional requirements. Very rapidly changing operational parameters caused by driver action require an exactly timed correlation of parameters such as fuel quantity and ignition timing.

3.1 Gasoline engines

3.1.1 External mixture preparation

External mixture preparation or mixture formation is the standard process used for spark-ignition engines. Internal mixture preparation, however, is desirable with regard to fuel consumption (see Chapter 3.1.2) as this represents a considerable fuel economy potential of approx. 20% at part throttle [3.75].

3.1.1.1 Combustion chamber shape, layout and compression ratio

To minimize HC emissions, compact combustion chambers with minimum gaps between combustion chamber surface and piston top are desirable since this reduces flame quenching although it entails an increase of NO_x emissions. Designs of this type provide small surface-to-volume ratios. The smaller the specific surface, the smaller is the surface of the cold cylinder walls that may cause the ignition flame to be extinguished. A favorable surface-to-volume ratio, i.e. a relatively small combustion chamber surface, also reduces emissions of unburned hydrocarbons (*Fig. 3.1*).

Optimized squish areas inside the combustion chamber are another important feature. They cause surfaces to approach each other during the upward piston stroke in such a way that displacement of the mixture results in optimized mixture movement. This promotes turbulent flow with intensive mixing and also introduces the HC particulates detached from the wall surface into a further oxidation process. The combustion chamber should be shaped in such a manner that an optimum charge movement is achieved.

The shape of the combustion chamber also has a direct effect on the compression ratio. To optimize fuel economy, the compression ratio should be as high as possible (so as to ensure high thermal efficiency) (see *Fig. 3.22*, p. 35).

An increased compression ratio, however, also produces higher NO_x and HC emissions. Nitrogen oxide emissions are impaired by increased peak combustion chamber tempera-

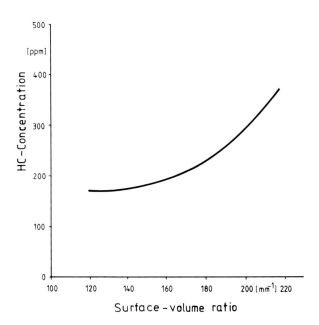

Fig. 3.1 Dependence of HC concentration on surface-to-volume ratio

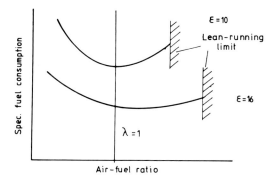

Fig. 3.2 Effect of compression ratio on lean operation limits

tures, whereas the increased fissuration of the combustion chamber (higher relative number of gaps) has a detrimental effect on unburned hydrocarbons. What is more, a higher compression ratio reduces exhaust temperatures due to improved efficiency. This, in turn, impairs post-reactions of unburned hydrocarbons and CO.

The improved suitability for leaner engine settings that is a result of improved ignition and combustion conditions, however, prevents HC emissions from increasing. This is a vital feature of lean-burn engines. Extremely high combustion ratios cannot be achieved in the full-load range, however, since the knock limit prevents ignition timing from being matched to efficiency in an optimum manner.

Figure 3.2 points out the basic effect of increasing the compression ratio. In addition to extending the operating range with regard to the air/fuel ratio, lean-burn operation also provides considerable fuel savings, a fact that in turn reduces CO_2 emissions.

The exhaust temperature reduction achieved as the compression ratio is raised may, however, have a detrimental impact on the light-off performance of the catalytic converter in the warming-up phase. Retarded ignition timing may offset this drawback.

As pointed out before, nitrogen oxide emissions are impaired by increased peak temperatures at higher compression ratios whereas unburned hydrocarbon emissions are impaired if the combustion chamber is extremely fissured (higher relative number of gaps). On the other hand, such designs respond better to lean-burn settings, although this positive effect on exhaust emissions cannot be exploited directly in systems based on $\lambda = 1$ control.

In systems with $\lambda = 1$ control, however, the EGR rate may be raised at higher compression ratios since the misfire limit is moved further towards the "leaner" region. This, in turn, reduces NO_x emissions and, as an additional benefit, improves specific fuel consumption.

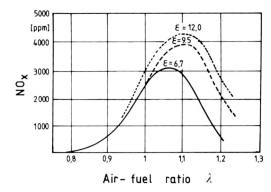

Fig. 3.3 Effect of compression ratio on nitrogen oxide concentration

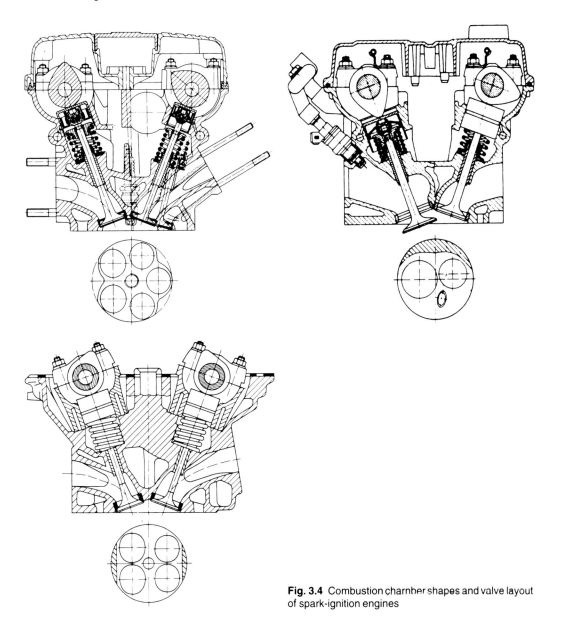

Fig. 3.4 Combustion chamber shapes and valve layout of spark-ignition engines

Figure 3.3 shows the NO_x concentration as a function of the air/fuel ratio for various compression ratios.

The maximum NO_x concentrations increase along with the compression ratio and are shifted towards higher excess air ratios at higher compression ratios. This is explained by the higher temperature level.

Exhaust emission characteristics are also influenced by the spark plug position, number of valves and valve gear layout, e.g. variable valve timing. Four or five-valve engines allow the spark plug to be positioned centrally. A compact combustion chamber with a central spark plug position that allows combustion time to be kept as short as possible thanks to short combustion paths and increased charge movement has a positive effect on HC emissions. Combustion noise constraints, however, tend to limit the speed of flame

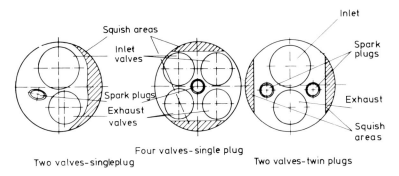

Fig. 3.5 Location of squish areas and spark plugs

propagation in the mixture. More rapid flame propagation can be enhanced by optimum sizing and location of squish areas (e.g. 10 to 15% of the piston area) that intensify this charge movement. In two-valve engines, flame propagation can be accelerated by using two spark plugs per combustion chamber (*Fig. 3.5*). *Figure 3.4* shows some common combustion chamber shapes and valve arrangements.

The combustion chambers of today's spark-ignition engines are usually located in the cylinder head, as opposed to direct-injection diesel engines where the piston incorporates the combustion chamber.

Pistons of spark-ignition engines either have a slight recess or are flat-topped. Pent-roof combustion chambers have proved advantageous on multi-valve engines. Valve pockets in the piston tops are detrimental to HC emissions as they contribute to an excessively fissured combustion chamber shape.

Figure 3.5 shows some possible designs of spark plug position, squish areas and valve layout, using two-valve and four-valve designs as examples.

To improve mixture induction in the lower part-load range, i.e. at low mass flow rates, one inlet port of multi-valve engines can be designed as a swirl port. The second port is activated at higher loads and engine speeds and is then used as a charging port.

3.1.1.2 Mixture preparation, mixture control and direct fuel injection systems

Both in spark-ignition and in diesel engine combustion processes, it is not possible to minimize all exhaust components simultaneously. As pointed out before, only carbon monoxide can be influenced almost exclusively via the λ coefficient if secondary parameters such as charge stratification, bore-to-stroke ratio etc. are disregarded. NO_x and HC emissions, however, are subject to influences by a number of other parameters. Mixture formation is therefore of paramount importance as far as pollutant emissions are concerned.

Due to the narrow misfire limits of the mixture, spark-ignition engines will tolerate only limited variations of the air/fuel ratio. A λ range with slight excess air offers advantages for low HC and CO emissions and low fuel consumption. Nitrogen oxides, however, are at their maximum here. In the $\lambda < 1$ range, oxygen deficiency causes a significant CO increase, and in many cases the misfire limit is already reached at λ values > 1.2, therefore causing HC emissions, in particular, to increase sharply.

Increasingly stringent emission standards require further improvement of mixture formation and mixture preparation systems [3.64] [3.75]. Parameters such as start of injection and, hence, preinjection angle (before the inlet valve opens), creating an air envelope around the injection jet, multi-point injection etc. are only some of the measures adopted to improve operation characteristics. As an example of measures, *Fig. 3.6* shows

3.1 Gasoline engines 25

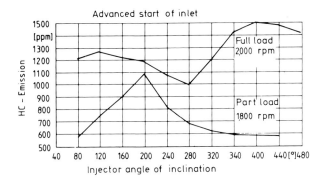

Fig. 3.6 Reduced HC emissions achieved by modifying the pre-injection angle [3.48]

how variations of the fuel pre-injection angle affect HC emissions [3.48]. In this example, HC emissions are low at large pre-injection angles in the part-load range. At full throttle, minimum HC emissions are achieved at different angles. Engine management systems may be used to obtain the optimum start of injection for any load-rpm range.

Figure 3.7 [3.49] shows how an injection nozzle surrounded by an air envelope at transient operating conditions affects HC emissions in the initial phase of the FTP testing cycle.

Lower HC emissions achieved by improved mixture preparation also improve light-off conditions of the catalytic converter and therefore yield higher conversion efficiencies in the initial test phase.

Map-controlled cylinder-selective injection systems (both for load and rpm) are often essential to ensure stable mixture preparation that complies with exhaust emission standards. The term "cylinder-selective" means that the mixture supplied to each engine cylinder is adjusted individually and is governed by a variety of operating and ambient conditions.

Experiments involving simultaneous and sequential injection (*Fig. 3.8*) show that a suitable correlation of injection timing relative to the opening timing of the inlet valve has a positive effect on lean-burn characteristics and, therefore, on HC and NO_x emissions [3.2]. In the case of simultaneous injection, fuel is injected at the same moment in all cylinders. As a result, the residence time available for mixture preparation in the intake rail

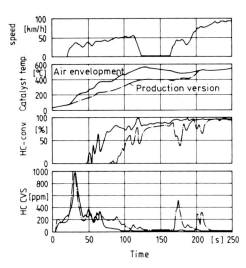

Fig. 3.7 Effect of air envelopes around injection nozzles on the initial phase of the FTP cold test [3.49]

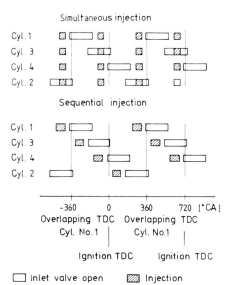

Fig. 3.8 Simultaneous and sequential injection [3.2]

differs from one cylinder to another. Other important operating conditions, e.g. acceleration, overrunning and warming-up, are also impaired by simultaneous injection.

Optimum mixture formation is of particular importance under transient conditions [3.77]. Good dynamic behavior is achieved by minimizing wall film formation and by synchronizing injection to the intake process during the opening period of the inlet valves. This, however, requires additional measures such as creating air envelopes around injection valves and inducing a corresponding charge motion to compensate for the resulting lack of homogeneity.

Further improvements of mixture stability and, hence, of emission characteristics, may be achieved by using lambda (λ) control systems with three-way catalytic converters. A λ sensor located ahead of the catalytic converter registers the air/fuel ratio and corrects the amount of fuel across related actuators. The system operates in a very narrow range near $\lambda = 1$ (refer to Chapter 5).

Modern engine management systems meet the above mentioned fuel system requirements. Important parameters include optimum adaptation of the mixture to a variety of operating conditions and equal distribution of the mixture (λ and cylinder charge) to the individual cylinders.

To ensure correct air supply, it must mad sure that the intake resistances of all cylinders are equal and that symmetrical vibration characteristics of the mixture column in the inlet system are achieved in order to provide each cylinder with an equal amount of charge. Similar requirements apply to the exhaust system. If $\lambda = 1$ control is to be ensured for each cylinder even at uneven cylinder charges, cylinder-selective injection requires a separate oxygen sensor for each cylinder.

3.1.1.3 Ignition timing and spark plugs

Ignition timing and the ignition energy supplied as well as the shape and position of the spark plug have a significant effect on exhaust emissions. Retarding the ignition timing increases exhaust temperatures, thus creating favorable boundary conditions for combustion processes in the exhaust system and for post-reactions of HC and CO. Since the process peak temperature drops at the same time, NO_x emissions are lower. Retarding the ignition

3.1 Gasoline engines

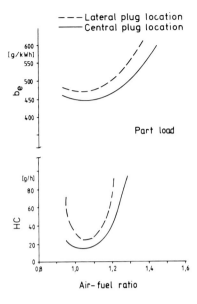

Fig. 3.9 Effects of spark plug position on HC and consumption [3.3]

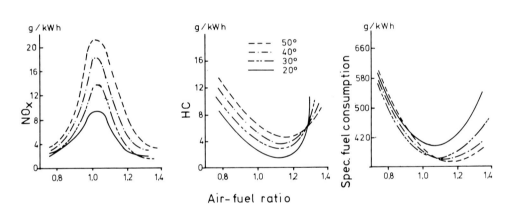

Fig. 3.10 Impact of air/fuel ratio and ignition timing on pollutant emissions [3.3]

timing, however, impairs fuel consumption, and therefore causes CO_2 emissions to increase again. On the other hand, retarded ignition settings are an effective means of shortening the warming-up phase and ensuring early light-off of the catalytic converter. Increasing the ignition energy has no significant impact on exhaust characteristics, although it allows the operating limits to be shifted towards lean mixtures, resulting in the known benefits on NO_x emissions. The exhaust recirculation rate can in this way be increased on engines with $\lambda = 1$ control.

To ensure optimum combustion of the air-fuel mixture, a large volume between spark plug and piston crown is required. A greater distance between spark plug electrode and piston crown therefore offers advantages. A central spark plug position giving short flame paths, as found on engines with four and five valves per cylinder, results in compact combustion chambers and high conversion speeds and therefore in low HC emissions and reduced fuel consumption, cf. *Fig. 3.9* [3.3]. An opposite effect on HC emissions is observed at the higher compression ratios achievable in multi-valve engines. In addition, increasing conversion speeds also lead to increased combustion noise.

If two spark plugs are used in two-valve engines, the flame paths can be shortened (as in multi-valve engines), although at the cost of impaired nitrogen oxide emission levels.

The influence of ignition timing on NO_x and HC emissions as well as on fuel economy is highlighted in *Fig. 3.10*. Advancing the ignition timing produces increased nitrogen oxide emissions across the entire λ range. In the stoichiometric range, in particular, advancing the ignition timing results in an excessive increase of NO_x emissions. HC emissions show similar tendencies. Since the exhaust temperatures decrease when ignition timing is advanced, post-reactions in the exhaust system are reduced and HC emissions increase. Based on the fact that the measures indicated improve lean operation characteristics and that, at the same time, the constant-volume process giving optimum efficiency is approached, fuel consumption is appreciably lower in the lean range. Knock characteristics prevent the optimum ignition timing from being matched to the effective efficiency, especially at full throttle and at high compression ratios. As a compromise, retarding the ignition timing is often necessary.

3.1.1.4 Exhaust gas recirculation (EGR)

Exhaust gas recirculation (EGR) refers to the introduction of exhaust gas into the fresh charge or the inlet air, respectively. A distinction is made among two types of EGR:

- External exhaust gas recirculation. It is achieved by feeding the exhaust gas externally from the exhaust system into the intake system.
- Internal exhaust gas recirculation. It is achieved by valve overlap and, hence, impaired cylinder scavenging. The residual gas content, i.e. the share of burned gases in the cylinder is increased.

The main goal of exhaust gas recirculation is to reduce NO_x emissions. This is essentially accomplished by the following factors:

- The heat capacity of the recirculated exhaust gas is higher than that of air. This reduces the temperature increase if the heat quantity released during combustion remains equal
- Reduction of O_2 partial pressure, resulting in a reduced oxygen mass in the cylinder since part of the combustion air is replaced by exhaust gas of lower oxygen content
- Reduction of combustion speed, resulting in an additional temperature reduction

To reduce NO_x emissions even further, the exhaust gas may be cooled. A temperature reduction of approx. 50% may be achieved if specially designed coolers are used.

Internal exhaust gas recirculation makes use of valve overlap during the charge changing process (see Chapter 3.1.1.5). Quantitative monitoring or control of the recirculated content by making use of valve overlap is only possible to a limited extent. One input possibility is using phase converters to rotate the inlet camshaft with regard to the exhaust camshaft, thus modifying valve overlap. Variable valve drives give even better results.

Excessive residual gas contents, especially near the idle range, cause misfire and rough engine running and produce a corresponding increase of unburned hydrocarbon emissions.

External exhaust gas recirculation is based on adding exhaust gas to the intake system. The air-fuel mixture and the added exhaust gas are drawn in by the engine. The use of EGR systems, however, only makes sense in the part-load range since the full-load range is operated at $\lambda < 1$ for performance reasons, and NO_x concentrations therefore are lower in this range. What is more, exhaust gas recirculation at full throttle would restrict engine output.

3.1 Gasoline engines

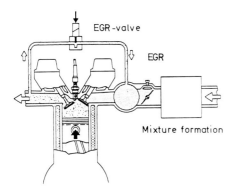

Fig. 3.11 Schematic design of an exhaust recirculation system [3.3]

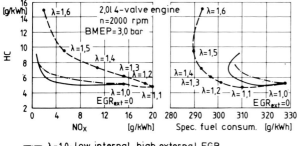

— · — λ=1,0; low internal, high external EGR
——— λ=1,0; low external, high internal EGR
– – – Lean operation, external EGR=0

Fig. 3.12 Effect of exhaust gas recirculation and lean-burn operation on pollutant emissions and fuel economy [3.8]

Figure 3.11 shows the basic layout of an exhaust gas recirculation system used with a carburetor or single-point injection system.

The EGR frequency valve usually is a diaphragm valve that is timed electrically to release or block a desired time section, feeding a metered quantity of exhaust gas into the intake system. The valve may be map-controlled. A common exhaust gas recirculation rate is 5 to 10% (5 to 10% of the total inducted mixture consist of exhaust gas). Depending on the degree of dilution of the charge, NO_x emissions may be reduced drastically in this way (*Fig. 3.12*), although at the cost of increased HC emissions. These may be kept virtually constant across a wide range in the part-load range if the engine is operated at stoichiometric mixture settings. A reduction of specific fuel consumption is also possible.

Engines with improved mixture formation (rich cloud in spark plug area) can also be operated at far higher exhaust gas recirculation rates of up to 20%, offering potential fuel consumption savings of up to 7% and lowering HC and NO_x pollutant emissions by approx. 35%. Today, exhaust gas recirculation is used by many engine concepts to comply with emission standards.

3.1.1.5 Valve timing

The valve timing arrangement is characterized by conflicting targets. High torque requirements at low engine rpm and high maximum output with low fuel consumption, good idle characteristics and low raw emissions are the major targets that cannot be met with fixed valve timing. Systems that provide variable timing (especially of the inlet valves) have

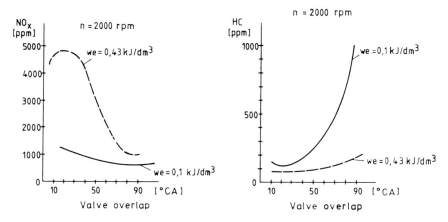

Fig. 3.13 Effect of valve overlap on pollutant emissions

therefore found widespread use in recent years. In this context, shifting the inlet timing has a relatively greater effect on the above engine parameters than any modifications to the exhaust timing.

Modern engines make increased use of systems (e.g. phase converters) that allow the start of valve opening to be varied in accordance with the engine operating point [3.4] [3.7] [3.40].

Figure 3.13 shows, by way of example, the relationship between NO_x and HC for two load positions if valve overlap is modified. Other investigations have shown that larger valve overlap rates allow HC emissions to be reduced significantly, especially when the engine is cold [3.41].

Figure 3.14 shows the effect of an advanced or retarded start of inlet on HC emissions. Advancing the inlet opening timing will in this case yield low HC emissions at high loads. At low loads, retarding the inlet opening is preferable as a means of minimizing HC emissions.

Variable valve timing systems of this type may also help to stabilize the combustion process [3.23]. Camshaft adjustment offers several benefits:

- Raw emissions are reduced,
- Torque increases in the lower and medium rpm range without affecting power in the upper rpm range,
- Good idle stability is achieved.

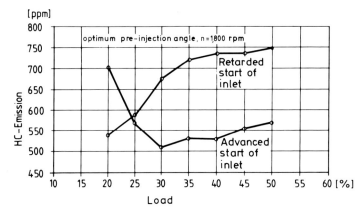

Fig. 3.14 HC reduction caused by camshaft adjustment [3.48]

3.1 Gasoline engines

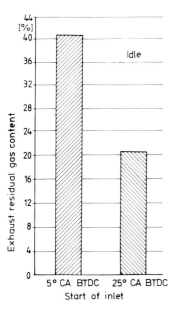

Fig. 3.15 Residual gas content at advanced and retarded timing [3.4]

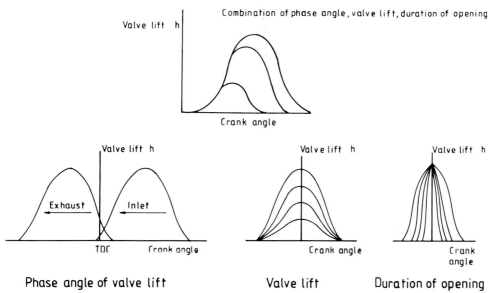

Fig. 3.16 Possible designs of variable valve timing

If high nominal engine power is a main objective, a longer opening period and large valve overlap (to improve cylinder charge by using the supercharging effect) are required.

At idle and near the idle range, valve overlap should be small in order to ensure stable idling, contributing at the same time to reduced HC emissions by lowering the residual gas content [3.24].

In those ranges where combustion stability is ensured, selecting valve overlap in a suitable way may help increase internal exhaust gas recirculation and will contribute to achieving reduced NO_x emissions.

Figure 3.15 shows how residual gas contents differ at advanced and retarded idle settings.

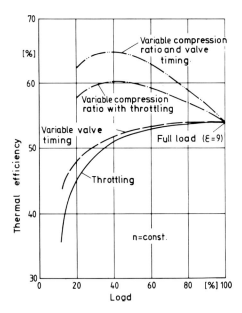
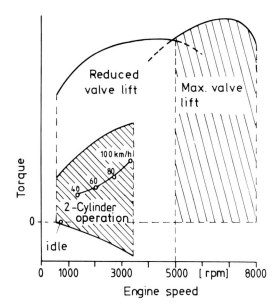

Fig. 3.17 Basic effect of variable valve timing and variable compression ratio on thermodynamic efficiency [3.56]

Fig. 3.18 Load control by variation of valve lift and cylinder cut-off [3.54]

Figure 3.16 shows some potential ways of utilizing valve timing. In addition to varying the phase angle, changed valve lift and valve opening timing or a combination of those possibilities allow both emission levels and engine efficiency and, hence, fuel economy to be influenced [3.25]. Systems of this type also allow alternate load control systems, turbulent control and cylinder cutout to be implemented [3.45] [3.50] [3.51] [3.52] [3.54].

Variable valve timing systems can also be used to reduce fuel consumption. *Figure 3.17* shows the basic influence of variable valve timing and variable compression ratios on thermodynamic efficiency [3.56]. The bottom part-load range is one particular area where high efficiencies can be achieved by reducing throttling losses.

Figure 3.18 shows the potential of load control that is obtained e.g. by modifying valve lift. The engine is operated with reduced valve lift at engine speeds of up to 5,000 rpm, and with increased valve lift beyond this speed. Valve timing is modified accordingly. In the lower load-speed range and, to a certain extent, in the part-load range, two cylinders are deactivated across the inlet valves. Cutting down on load-change losses and cutting off cylinders yields fuel economy improvements of approx. 10% according to the Japan 10–15-mode test cycles [3.54]. Research on engines that use variable valve lift settings in order to achieve a load control effect has revealed similarly improved fuel economy [3.53]. Additionally, NO_x emissions can be reduced particularly well in the lower load range as inlet closes earlier in this range and the resulting gas temperatures are therefore lower. HC emissions also are lower.

3.1.1.6 Inlet port configuration and swirl

Arrangement and shape of the inlet ports have a significant effect on the combustion process and, hence, on emission characteristics [3.66]. Of equal importance is the intake

3.1 Gasoline engines 33

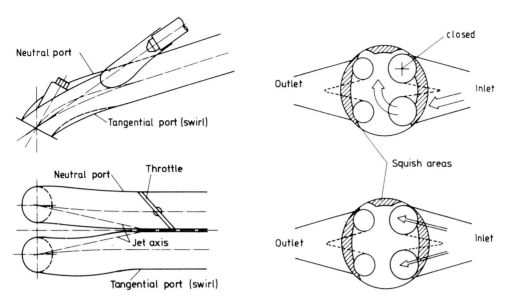

Fig. 3.19 Inlet port configuration (schematic) [3.2]

Fig. 3.20 Port cutout controlled by inlet valve

pipe design as this allows the above parameters to be influenced to an even greater degree. This influence focuses essentially on induction characteristics and therefore on mixture preparation and is governed primarily by the below parameters:

- Inlet pipe length and inlet pipe diameter,
- Number of intake pipes (ports) per cylinder,
- Position and number of injection nozzles,
- Shape and length of inlet ports,
- Port cutout used to increase inlet speed and swirl generation in multi-valve engines,
- Injection angle of injection jet.

Inlet swirl can be intensified using specially shaped ports. However, this usually entails charge losses and impaired HC emissions, especially if mixture formation is less than perfect. One way of avoiding this problem of charge losses is to adopt a multi-valve design, designing one port as a swirl port and the other(s) as a charging port.

Figure 3.19 shows a schematic overview of one possible configuration. In this design, the charging port is cut off by a throttle valve in the lower load and rpm range. Activating and deactivating one of the inlet valves is an even more efficient solution, cf. *Fig. 3.20* [3.26].

If at least two inlet valves are used, a roller-shaped tumble instead of a swirl effect can be generated at right angles to the cylinder axis. The rotating axis of the tumble is located perpendicular or at an oblique angle to the cylinder axis. As with swirl, the tumble effect increases turbulence and, hence, combustion speed. This significant acceleration of combustion has a stabilizing effect and improves tolerance with regard to charge dilution. In systems with low charge movement, though, charge dilution means that cyclic oscillations increase significantly.

Additional mixture stratification near the spark plug ("rich cloud") is also possible, allowing the exhaust gas recirculation rate to be increased further. The charge movement is of greater influence than mixture stratification.

3.1.1.7 Bore-stroke relation and cylinder volume

The bore/stroke ratio and the cylinder volume also are parameters having an influence on exhaust emissions and on fuel economy. *Figure 3.21* shows the dependence of relative HC emissions on these parameters. The longer the stroke relative to the engine bore, the lower are HC emissions and part-load fuel consumption. These relationships cannot always be selected freely, however, since criteria such as mass forces, combustion chamber design, installation space, available production equipment etc. permit only small variations for passenger vehicle use.

Varying the bore/stroke ratio essentially allows the surface/volume ratio parameter to be influenced. This ratio, i.e. the combustion chamber surface, is higher on a short-stroke engine and therefore is less favorable than on a long-stroke engine. Additional important factors are the compression ratio, the change of the gas temperatures towards the end of the combustion and, consequently, the post-reaction of hydrocarbons [3.27] [3.28].

Figure 3.22 shows the dependence of nitrogen oxide concentrations on a variety of parameters. On the whole, long-stroke engines offer a better potential for improving efficiency, especially in conjunction with 4-valve technology [3.35]. The underlying reason is that the efficiency gains at higher compression ratios may be higher in longer-stroke engines than in short-stroke engines. This is due to the less favorable combustion chamber shape that is typical for shorter engine strokes.

As is to be expected, long-stroke engines compare rather less favorably when NO_x concentrations are considered since the compact combustion chamber yields higher combustion temperatures. Larger swept volumes (displacements), in particular, also emit higher nitrogen oxide concentrations in the exhaust [3.36]. On the other hand, unburned hydrocarbons show opposite characteristics.

Numerous research projects have shown that a displacement of 450 to 500 c.c. is the optimum in terms of fuel economy.

To conclude this section, *Fig. 3.23* shows the dependence of the above parameters on internal efficiency and air/fuel ratio [3.36]. Long-stroke engines also offer advantages with regard to suitability for leaner settings. The higher internal efficiency achievable yields

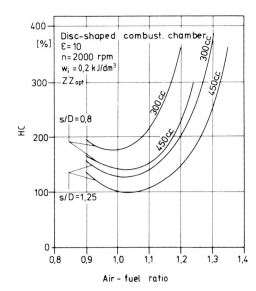

Fig. 3.21 HC emissions as a function of air/fuel ratio [3.28] [3.36]

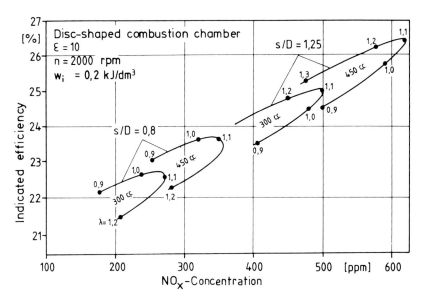

Fig. 3.22 Efficiency and nitrogen oxide emissions [3.28] [3.36]

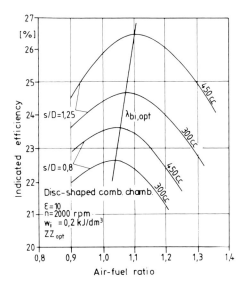

Fig. 3.23 Effect of s/D and swept volume on internal efficiency [3.28] [3.36]

higher torque at lower engine speeds. As this allows for longer transmission ratios, the fuel consumption can be reduced even further.

3.1.1.8 Cooling system

One way of reducing HC and NO_x emissions while at the same time optimizing the lean-running properties of engines not operating with $\lambda = 1$ control is to increase component temperatures in the part-throttle range.

This improves mixture preparation and ignition conditions and thus allows, among others, higher exhaust recirculation rates to be achieved in concepts based on $\lambda = 1$ control.

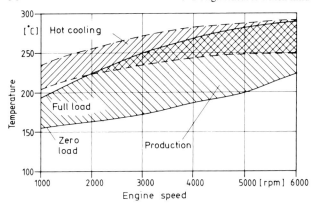

Fig. 3.24 Component temperature at the upper ring reversal point on the cylinder wall [3.78 [3.80]

Higher component temperatures may be obtained e.g. if cylinder head and cylinder block are cooled separately, or by adopting hot or evaporation cooling which is a particularly efficient method. A disadvantage, however, is that the engine is more susceptible to knocking and may suffer from power losses. *Figure 3.24* shows the effect of modified cooling conditions on component temperature.

The suitability for lean settings thus achieved also reduces pollutant emissions by a considerable margin. This is evident from *Figs. 3.25* and *3.26*, with *Fig. 3.26* showing the results obtained for hot cooling and turbulence generation.

The fuel economy achievable with hot or evaporation cooling is 5 to 7% for $\lambda = 1$ control concepts.

Coolant temperature and, as a result, combustion chamber temperature have a pronounced effect on HC emissions that are generated when the flame is quenched at the cold cylinder wall.

Low temperatures along the combustion chamber walls cause HC emissions to increase dramatically to an extent that cannot even be compensated by retarding ignition timing. Measures to achieve faster heating of the coolant are therefore required. This may be accomplished, to name some examples, by using a latent heat accumulator, introducing external heat or by reducing heat capacities.

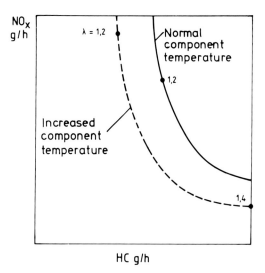

Fig. 3.25 Effect of component temperature on NO_x and HC [3.80]

3.1 Gasoline engines

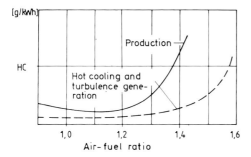

Fig. 3.26 Effect of hot cooling and turbulence generation on HC emissions [Source: VW]

3.1.2 Lean burn engine concepts

A three-way closed-loop catalytic converter currently constitutes the most efficient way of reducing pollutant emissions of spark-ignition engines. The drawbacks of such a system, especially in terms of fuel economy, design complexity and costs, have led to a desire for alternative solutions. Two areas have recently shown progress:

- Lean burn concepts
- Alternative fuels

Both design trends may be combined with oxidation and reduction catalytic converters, thus allowing emissions to be reduced even further.

If we take a closer look at the curve of the major exhaust components, it is evident that the NO_x and CO components are reduced significantly when operating the engine at $\lambda \gg 1$. Only unburned hydrocarbons increase as the air/fuel ratio is raised since we now approach the misfire limits of the air-fuel mixture. *Figure 3.27* shows the basic curves of the concentrations of major pollutants beyond the operating range of a conventional engine.

The lean burn concept focuses on reducing raw emissions by shifting the operating point until emission standards can be complied with. This may be accomplished for NO_x and CO emissions by moving the operating point into lean-burn regions. The resulting increase of HC concentrations, however, is a definite disadvantage. The advantage of reducing specific fuel consumption is, in some cases, accompanied by the drawback of inadmissible torque variations.

If the operating limits of the spark-ignition engine can be moved towards far higher air/fuel ratios without impairing operating characteristics, the following advantages may be exploited:

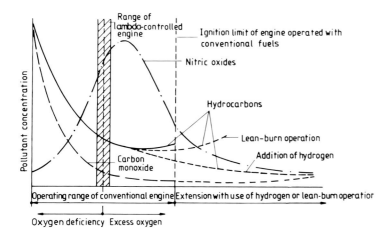

Fig. 3.27 Pollutant concentrations emitted when extending the engine operating range (e.g. lean-burn operation)

- Low fuel consumption and therefore lower CO_2 emissions,
- Low CO and NO_x emissions.

Figure 3.28 shows the results obtained for two conventional engines of different compression ratios fitted with catalytic converters compared to those of a lean-burn engine without catalytic converter.

The emission characteristics must be influenced in such a way that the emissions remain below the legal limits. Lean-burn engines thus show the following results:

- *CO:* Thanks to lean-burn operation, CO emissions are relatively low.
- *HC, NO_x:* Both pollutants have to be minimized.
- *Fuel consumption:* Thanks to lean-burn operation, fuel economy improves by approx. 10 to 15%.
- *CO_2:* Lower fuel consumption means that CO_2 emissions are also reduced by 10 to 15%.

Pollutant emissions of two basic engine versions, i.e. lean-burn engines and $\lambda = 1$ designs, used on a medium-sized vehicle without catalytic converter are compared with the U.S. standards in *Fig. 3.29*.

When nitrogen oxide emissions are plotted as a function of unburned hydrocarbon emissions, a curve like the one shown in *Fig. 3.30* may result for a given engine. This

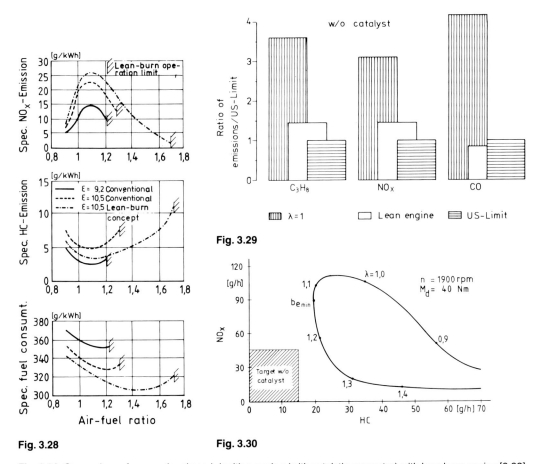

Fig. 3.28 **Fig. 3.30**

Fig. 3.28 Comparison of conventional spark-ignition engine (with catalytic converter) with lean burn engine [3.23]
Fig. 3.29 Comparison of raw emissions measured in FTP-75 testing [Source: VW]
Fig. 3.30 HC–NO_x map of a lean-burn engine without catalytic converter [3.80]

diagram clearly indicates the effect of the air/fuel ratio. As is evident from the curve, lean burn operation reduces NO_x emissions dramatically whereas HC emissions increase sharply at the same time. This curve does not have a common minimum for both values. At $\lambda = 1$, NO_x emissions still are high. These, however, can be reduced by fitting a three-way catalytic converter.

Within the range of low NO_x emissions, an oxidation catalyst is required to reduce the HC peaks.

If it is assumed that lean burn operation at $\lambda = 1.2$ to 1.4 can be accomplished in a wide range of the map, the boundary curve of *Fig. 3.30* shows that the target region to be reached in order to meet the legal requirements without using a catalytic converter is still a long distance away.

The air/fuel ratios accomplished with real lean-burn engines are indicated in *Fig. 3.31*. These engines achieve values λ values up to 1.5 at part throttle. At full throttle, the λ value is set to <1 in order to maximize output, whereas values approximating the stoichiometric ratio are adjusted near idle.

A decisive factor of such concepts is how the lean operating potential of an engine can be improved. If the operating range of a lean burn engine is subdivided into several segments, the following statements can be made:

- *Up to $\lambda = 1.3$*

Simple measures are sufficient to ensure stable engine operation. Stabilizing the combustion is the major target here. Cyclic variations from one operating cycle to another or among the individual cylinders have to be reduced by improving mixture distribution.

An increased compression ratio incorporating knock control and squish flow characteristics that create a high degree of turbulence is required. This may also be accomplished by charge motion generated in the inlet.

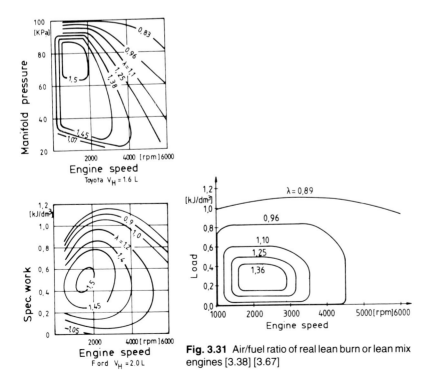

Fig. 3.31 Air/fuel ratio of real lean burn or lean mix engines [3.38] [3.67]

The residual gas content should be kept as low as possible by providing a corresponding valve overlap. In this air/fuel ratio range, NO_x emissions continue to be comparatively high. For this reason, it is desirable to move the operating range further towards the lean range.

- *Up to $\lambda = 1.8$*

To stabilize combustion in this lambda range, additional complicated measures (e.g. a defined charge motion) are required since HC emissions would increase significantly if no additional measures were introduced. This is due to the fact that the mixture may not ignite any more or that the combustion duration may be extremely long (*Fig. 3.32*).

The extended combustion duration caused by low flame speeds means that the fuel has not yet been fully combusted when the exhaust valve opens. This causes some of the unburned fuel to escape into the exhaust port and results in a corresponding increase of HC emissions. Excessive combustion duration also impairs fuel consumption as the engine processes continue to move away from constant-volume combustion conditions that would yield optimum efficiency. *Figure 3.33* shows the combustion speed curve versus the air/fuel ratio.

Present-day production engines already reach λ values of up to 1.8 [3.26]. NO_x levels are the same as those of engines with three-way closed-loop catalytic converters [3.23]. Fuel economy and, hence, CO_2 emissions are about 10 to 15% below those of engines operated at $\lambda = 1$.

To maintain stable combustion in this λ range, advanced features such as multi-jet Injection, special inlet port shaping, sequential injection, strategies for stratified charge and transient operation, swirl control and tumble are required [3.45] [3.46] [3.47].

Lean mix engines: A lean burn concept does not necessarily mean that the engine must be operated at settings of $\lambda > 1$ across the entire operating range. Certain sections of the map can in fact be excluded from extreme lean burn settings. This is referred to as lean mix engines. Mixture metering is achieved as follows:

- Idle: preferably $\lambda > 1$,
- Part throttle: lean control to $\lambda \gg 1$,
- Full throttle: $\lambda < 1$ or $\lambda = 1$ to achieve optimum engine output.

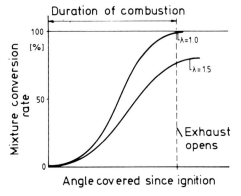

Fig. 3.32 Schematic diagram of combustion duration and maximum conversion rate of lean-burn mixtures

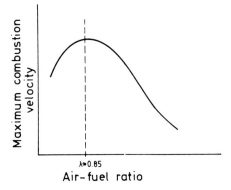

Fig. 3.33 Combustion speed as a function of air/fuel ratio

3.1 Gasoline engines

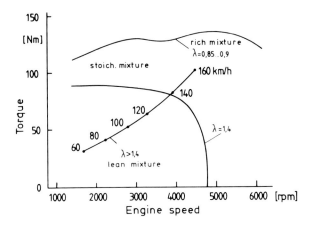

Fig. 3.34 λ ranges of a lean-mix engine [3.55]

Tuning of production engines [3.47] shows that idle and part-throttle settings may be adjusted to give a lean mixture for optimum fuel economy. A stable idle is of critical importance here. During acceleration, the engine is supplied with a stoichiometric or rich air-fuel mixture. The high NO_x emissions produced in the process are converted by a three-way catalytic converter. Since nitrogen oxides are not converted by the catalytic converter in the lean range, excess air ratios must be as high as possible in order to achieve low NO_x emissions. *Figure 3.34* shows some potential ranges of λ tuning [3.47].

Lean operation at settings $\lambda > 1.4$ covers a wide load and rpm range. The fuel economy achieved with this type of tuning amounts to approx. 9% for a real engine tested in the Japan 10–15 test cycle [3.55].

3.1.2.1 Lean burn management systems

When designing lean burn engine management systems, two approaches may be adopted:

- Lean-burn operation throughout, or
- Lean-mix operation.

As described above, the latter is an operational mode that, depending on requirements, covers a map with settings of $\lambda > 1$, $\lambda = 1$ and $\lambda < 1$. When designing a lean burn engine or the lean map range of a lean mix engine, two basic criteria have to be taken into account:

- Operate engine in a specified λ range in the lean-burn range to keep NO_x emissions low.
- Avoid exceeding a certain level of rough running or cyclic variations so as to ensure correct drivability and to avoid increase of HC emissions.

If the above criteria are taken into account, a λ range results that must be adhered to by implementing suitable controls. *Figure 3.35* shows a schematic diagram of this λ range and the resulting NO_x reduction.

As pointed out above, different strategies must be used for different map ranges to achieve optium tuning of an engine. Since the above ranges must blend with each other smoothly without causing drivability problems, a combination of quality and quantity management is required, i.e. certain operational ranges are covered by modifying mixture composition whereas others are covered by throttling the mixture. This is due to the fact that a lean mixture at wide open throttle may produce less specific work than a rich mixture

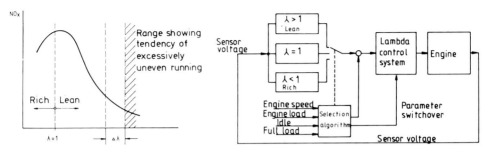

Fig. 3.35 Δλ range

Fig. 3.36 Block diagram of a λ range formation [3.29]

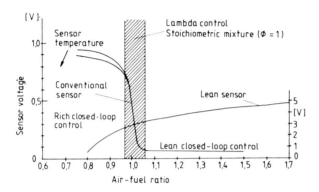

Fig. 3.37 Conventional lambda sensor and lean sensor maps

at partly closed throttle positions [3.29]. *Figure 3.36* shows one way of generating a suitable λ range ($\lambda < 1$, $\lambda = 1$, $\lambda > 1$) with this type of control.

To ensure efficient fuel metering that matches the variable boundary conditions, a λ sensor (oxygen sensor) is required that is also operative in the lean range (lean sensor). An important aspect of this sensor is that the sensor output signal is not significantly dependent on the exhaust temperature in the lean range. Depending on the oxygen content of the exhaust, the lean sensor supplies a voltage signal that is used to control the mixture composition.

In the rich range, the additional dependence on exhaust gas temperature will also have to be compensated. When the sensor temperature rises, its internal resistance drops (logarithmically), providing an exact correlation of sensor voltage with air/fuel ratio. *Figure 3.37* shows the map of this type of sensor.

3.1.3 Internal mixture preparation

Internal mixture preparation of spark-ignition engines can be accomplished by injecting the fuel directly into the combustion chamber (high-pressure injection). Spark-ignition engines with internal mixture preparation offer a significant potential for reducing fuel consumption and for improving drivability dramatically while the engine is still cold if, as in the case of diesel engines, they are operated with quality management (without throttling device). Efficiency at part throttle can be improved by reducing charge-changing work and improving thermodynamic efficiency that is due to a

- *Higher effective compression ratio*

3.1 Gasoline engines

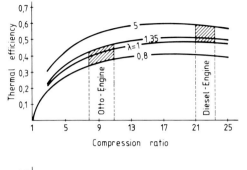

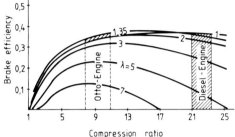

Fig. 3.38 Thermal and effective efficiency as a function of compression ratio

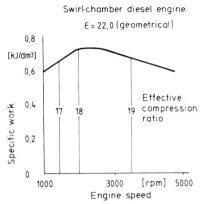

Fig. 3.39 Effective compression ratio of a diesel engine

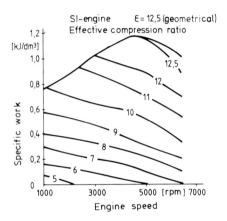

Fig. 3.40 Effective compression ratio of a spark-ignition engine

The higher the compression ratio, the higher is the effective efficiency in the range of particular interest to spark-ignition engines (cf. *Fig. 3.38*). In practice, the compression ratio is limited by the knock characteristics of the fuel.

In the case of quality management and depending on induction conditions, the maximum possible air quantity is drawn in across the entire load and rpm range so that the prevailing pressure and temperature levels at the end of the compression ratio are virtually equal to the geometric compression ratio. *Figure 3.39* shows this characteristic for a swirl-chamber diesel engine with a geometric compression ratio of $\varepsilon = 22$ and an effective compression ratio of $\varepsilon = 17$ to 19. The throttled engine, on the other hand, is never able to exploit the geometric compression ratio to its full effect at part throttle and therefore yields only poor efficiency (*Fig. 3.40*). As is evident from this figure, the actual compression ratios under throttle control at part throttle remain far below the geometric compression ratio.

Any measure implemented to increase the effective compression ratio will reduce fuel consumption as a direct result if the other conditions remain the same.

- *Higher final compression temperature resulting from higher compression ratio*

In the case of quality management, the final compression temperature at a fixed compression ratio is only dependent on the engine rpm in an initial approximation and will

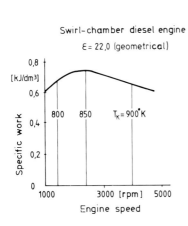 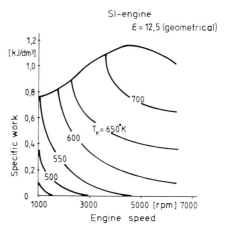

Fig. 3.41 Final compression temperatures of a swirl-chamber diesel engine

Fig. 3.42 Final compression temperature of a spark-ignition engine

therefore always attain relatively high levels. *Figure 3.41* shows this effect, taking a swirl-chamber diesel engine as an example.

The full-load combustion process of an engine with throttle control starts on a high thermal level whereas the temperatures present in the lower part-load range at the beginning of the ignition process are as low as 450 °K (cf. *Fig. 3.42*). This has a detrimental effect on ignition, combustion and lean-burn potential of the air-fuel mixture.

An essential requirement for quality management in internal mixture formation systems is stratification at part throttle. This means that the air-fuel mixture is enriched in the vicinity of the spark plug to ensure safe ignition yet the average combustion process can occur at an extremely lean mixture setting. This yields fuel consumption figures comparable to those of diesel engines as well as low NO_x and CO raw emissions. To reduce HC emissions, an oxidation catalyst has to be used. As shown in *Fig. 3.43*, part-load fuel consumption figures are reduced by 15 to 25% [3.12]. *Figure 3.44* shows the corresponding combustion chamber shape. The injection nozzle is a stepped pintle nozzle.

Due to the cycle used to obtain good fuel economy and low pollutant emission levels, the specific work levels of this direct-injection SI engine [3.12] are lower than those of spark-ignition engines with manifold multi-point injection. Supercharging may be used to increase output and specific work again. To obtain the power density of spark-ignition engines with external mixture formation, the engine has to be operated with a homogeneous mixture at full throttle. Pressure-atomizing direct-injection processes can be used for this purpose, e.g. in conjunction with air-distributing multi-hole nozzles, and injection timing can be varied accordingly. If injection timing is advanced by a comparatively large extent, any charge stratification that is present initially can be removed by charge motion and mixture homogeneity can be increased continuously up to the ignition point. Thanks to the increased residence time of the fuel in the combustion chamber, fuel preparation is relatively good.

An intensive charge stratification that is required for engine operation without throttling, however, means that ignition timing must be retarded further. While this ensures a high stratification level, it also means that fuel preparation is rather poor due to the reduced residence time of the fuel in the combustion chamber.

In addition to research on air-atomizing direct injection processes of SI engines using compressed air support for mixture preparation, increased effort is applied to developing mixture formation systems with gas support.

Design features

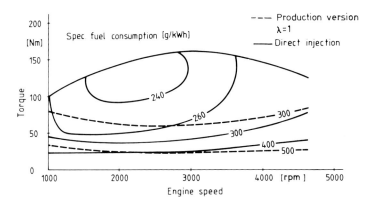

Fig. 3.43 Specific fuel consumption of a direct-injection spark-ignition engine with quality management [3.12]

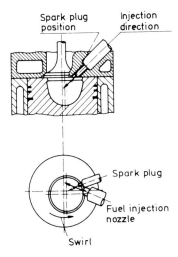

Fig. 3.44 Combustion chamber shape and injection nozzle position of a direct-injection spark-ignition engine [3.12]

Injection systems with compressed-air support allow the fuel to be introduced into the combustion chamber in a well-atomized condition, yet most of the fuel remains in a liquid state. Vaporization and pre-reactions of the fuel essentially occur inside the combustion chamber.

The injection energy can be obtained not only through external compressor output but also from the energy released in the cylinder itself during the compression and combustion process (cf. Chapter 4.1). To this effect, a section of the injector designed as an accumulator can be used not only to supply the injection pressure required but also to generate a temperature level sufficient to ensure fuel vaporization. Vaporization and partial pre-reaction of the fuel already occur inside the accumulator. When setting the injection timing, only the stratification requirements but not the mixture preparation requirements have to be taken into account (cf. Chapter 4.1).

Mixture formation systems of this type allow unthrottled engine operation down into the idle range. Whereas many pressure-atomizing direct-injection systems of spark-ignition engines can be operated without throttling at idle, this does not contribute to suitable emission concepts since HC emissions are very high. On the other hand, mixture formation systems using gas support as described above will also yield low HC emissions and excellent combustion stability at unthrottled idle operation (*Fig. 3.45*).

In the part-load range, NO_x emissions are the dominating characteristic to be taken into account in order to meet future emission limits with internal mixture formation

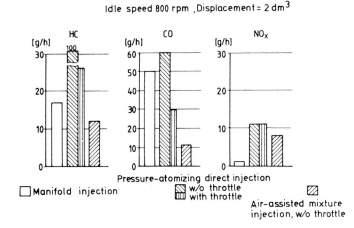

Fig. 3.45 Emissions of different mixture formation systems [Source: AVL]

systems. Again, combustion systems using gas support offer advantages over pressure-atomizing direct-injection systems.

Exhaust gas recirculation systems can be used to an increasing extent to reduce nitrogen oxide emissions as they offer comparatively low HC emissions and good EGR suitability. As a result, NO_x emissions are significantly lower than in direct-atomizing direct injection systems.

In certain areas (storing, long-term life), mixture formation systems of this type require higher system development effort than pressure-atomizing direct-injection systems.

As opposed both to lean-burn concepts utilizing external mixture formation and to pressure-atomizing mixture formation on direct-injection engines, gas-supported combustion systems of this type appear to be capable of achieving emissions below 0.4 g NO_x/mile according to FTP-75 testing with medium-sized vehicles. Systems of this type are currently in the predevelopment stage.

Figure 3.46 compares results of stationary measurements for three different injection systems at the representative part-load point of 2,000 rpm and an indicated specific work of 0.3 kJ/dm³. The comparison includes a multipoint manifold injection system as well as direct-injection and mixture injection systems.

Extremely lean mixture settings can be achieved with mixture injection, in particular, and will yield the known good fuel economy and emission levels. One important feature in

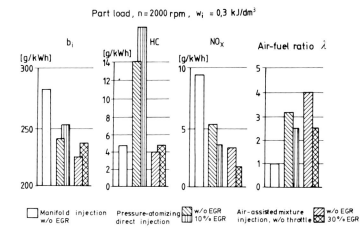

Fig. 3.46 Comparison of results obtained for different injection systems [Source: AVL]

this context is that HC emissions are particularly low. The part-load point shown in *Fig. 3.46* also shows operation with direct liquid injection without throttling the intake air. In contrast to this, fuel economy and exhaust emissions obtained with mixture injection are surprisingly good.

3.1.4 Gasoline engine potential in terms of emissions and fuel consumption

The potential of spark-ignition engines in terms of exhaust emissions and fuel economy has by no means been fully exploited.

Intensive research is being carried out on exhaust systems to comply with the ULEV limits (Chapter 8.4.2.1) for California. Major targets are to reduce raw emissions by implementing measures inside the engine, using alternative fuels or combinations of fuels, and to ensure fast lighting-off of the catalytic converter, e.g. by using external heating.

In addition, the lean-burn catalytic converter will be an important new feature as it allows NO_x emissions of lean-burn engines to be reduced significantly.

The use of alternate fuels, e.g. methanol, to reduce specific exhaust emission components is another element in the effort to cut down on exhaust emissions.

Reducing the warming-up period and improving preparation and adaptation of the fuel to the engine operating conditions (transient operation) are measures that keep down pollutant emissions while at the same time reducing fuel consumption. Other important elements in this respect include variable valve timing, variable compression ratios, cylinder cutout, direct injection, multi-valve technology, reduction of friction, "electronic" accelerators as well as supercharged engines that are not designed for maximum output but for minimum fuel consumption.

Further development of lean-burn and lean mix engines as well as two-stroke engines lead the way to further potential fuel savings. The measures that have so far proven to be effective will allow fuel consumption fuel to be reduced by more than 30% altogether.

The internal-combustion engine, however, is only one of several parameters that affect the vehicle. Medium-sized vehicles with fuel consumption figures of 3 to 4 l/100 km according to the Euromix formula can become a reality only if the entire vehicle is revised.

3.2 Diesel engines

Since load control of the diesel engine is achieved by modifying the air/fuel ratio, use of this parameter allows pollutant emissions to be influenced, in an initial approximation, only in the full-load range (as opposed to spark-ignition engines). Within this range, the injected quantity is limited to cut off fuel supply. The smoke or soot number determines the limits of this feature. Yet various additional possibilities are available to reduce exhaust emissions of diesel engines by implementing engine-related measures.

Formation and reduction of particulate matter and/or soot is of considerable importance on diesel engines.

As shown before, soot formation processes, in particular, involve local oxygen deficiency, dehydration and cracking reactions. Combustion temperature, mixture movement, fuel composition, diffusion speed, reaction time and quality of fuel atomization are some of the parameters that have a significant impact on the above components. The map of soot formation (soot emissions) versus injection timing (*Fig. 3.47*) shows that most of the soot formed can be reduced by post-oxidation [3.13].

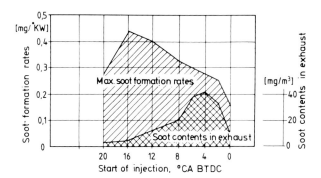

Fig. 3.47 Soot emissions vs injection timing [3.13]

3.2.1 Combustion process

Diesel engines are designed either as pre-chamber (indirect-injection or IDI) engines or as direct-injection engines, with pre-chamber engines being subdivided into swirl-chamber and precombustion engines. *Figure 3.48* shows a comparison of the three combustion processes using cylinder head sections as an illustration.

Most pre-chamber engine designs use the swirl-chamber process. Precombustion engines are currently manufactured by only one passenger car manufacturer. Direct-injection engines are dominated by the multi-jet process (also referred to as air-distributing process), as opposed to the single-jet process which is also referred to as the wall-adhesion or wall-distribution system.

The exhaust emissions and fuel consumption characteristics may be influenced to a significant extent by the selection of the combustion process. Pre-chamber type engines produce lower CO concentrations (emissions) than direct-injected engines since they achieve more intensive intermixing of fuel and air during the combustion process.

In a similar manner, chamber-type engines show lower NO_x and HC exhaust concentrations. *Figure 3.49* shows the characteristics for NO_x and CO. Direct-injection diesel engines offer a 15 to 20% bonus in fuel consumption and are therefore expected to gain increasing importance in the future since suitable means (e.g. exhaust aftertreatment) are available to reduce emissions appreciably.

The fuel consumption penalty of chamber-type engines results primarily from transfer losses and higher wall heat losses as the combustion chamber surface of this engine type is far larger (pre-chamber plus main combustion chamber).

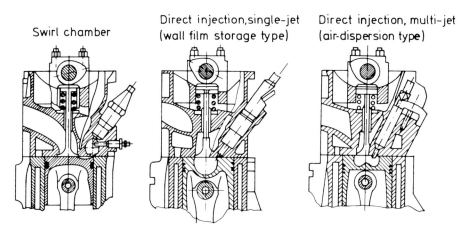

Fig. 3.48 Combustion processes of diesel engines [3.16] [3.63]

3.2 Diesel engines

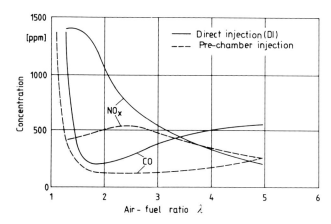

Fig. 3.49 CO and NO_x concentrations for Diesel engines [3.75]

Injection timing also has a significant impact on fuel consumption and pollutant emissions. The timing curve describes the injected fuel quantity relative to the crank angle. Start of injection and injection duration are the main parameters of the injection process.

These two parameters have to be adapted in a different way to the different diesel combustion processes. Precombustion engines, for example, require a longer injection duration than swirl-chamber engines. This duration is shortest on direct-injection engines.

In terms of minimizing fuel consumption and limiting emissions, an injection duration dependent on the engine operating point and matched to the start of injection is essential [3.3]. In addition to the above parameters, the injection pressure plays a decisive role. Whereas injection pressures in excess of 350 bar hardly offer any advantages at all on chamber-type engines, pressures far above 1,000 bar have been considered for direct-injection engines since they allow fuel consumption and emission characteristics to be improved. Pressures above 1,500 bar are desirable.

3.2.1.1 Pre-chamber engines

Two designs are of importance on chamber-type engines:

- Swirl-chamber engines
- Precombustion engines

At part throttle, the soot content in the diesel exhaust is higher than on direct-injection diesel engines. This is due to significant local air deficiencies inside the pre-chamber.

Part of the soot is oxidised further as the pre-chamber contents are fed into the main chamber. This, however, requires a certain temperature level that is not always reached at part throttle.

One goal of further development work on chamber-type engines in terms of emission levels is to optimize "feeding" of the pre-chamber with air and fuel and to optimize the design, e.g. shape, glow plug position, injection direction and swirl, transfer port and main combustion chamber shape.

Both versions, i.e. precombustion and swirl-chamber engines, are characterized by a two-stage combustion process. A rich mixture reacts inside the pre-chamber (air deficiency). A lean mixture is burnt in the main combustion chamber (excess air). Compared to single-stage combustion processes (direct injection), this type of combustion basically yields lower NO_x emissions. This is due to the fact that, while the chamber does reach high temperatures, the oxygen required for NO_x formation is not available. In the main combustion chamber, on the other hand, this situation is reversed.

a) Swirl-chamber engines: The swirl-chamber combustion process dominates in passenger vehicle diesel engines. In the top dead center position, the combustion chamber is separated into two chambers of roughly equal size interconnected across a passage. Fuel is injected into the swirl chamber filled with some of the combustion air. The eccentric location of the injection port in the swirl chamber produces a swirl movement inside the swirl chamber.

This process allows a two-stage combustion process to be used. In the same way as for the precombustion engine, several ways of modifying the engine characteristics are available to optimize the process. These are essentially:

- Compression ratio,
- Combustion chamber shape (contours and heat transmission),
- Main combustion chamber to swirl chamber volume ratio,
- Tuning of injection process (pressure, timing, duration, nozzle location etc.),
- Transfer port (length, diameter, angle) to main combustion chamber,
- Position and diameter of glow plug,
- Shape of main combustion chamber.

Further measures to influence engine characteristics are e.g. exhaust recirculation, supercharging and catalytic afterburning. An engine optimized according to the above criteria provides extremely low soot numbers within the characteristic map and thus ensures that virtually no visible components remain in the exhaust gas, cf. *Fig. 3.50* [3.15]. This also ensures acceptable fuel consumption figures, cf. *Fig. 3.51* [3.15].

Injection timing: Injection start and injection duration are two of the major factors that contribute significantly to the emission characteristics and, hence, to the emissions level of the essential pollutants, i.e. particulates, NO_x and HC. As the injection timing continues to be moved toward TDC, the combustion temperatures and, hence, nitrogen oxide emissions are lowered further. An added result is that the smoke number is lower.

The above advantages come at the expense of increased specific fuel consumption and HC emissions, as borne out by *Fig. 3.52* [3.14]. This increase of HC emissions is of

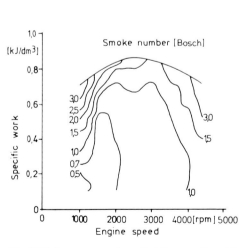

Fig. 3.50 Soot number characteristic map of a swirl-chamber engine

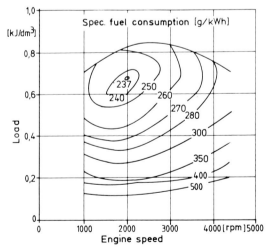

Fig. 3.51 Fuel consumption map of a swirl-chamber engine [3.15]

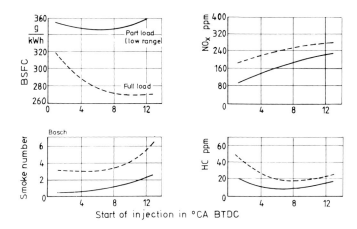

Fig. 3.52 Effect of injection timing on fuel consumption and emissions [3.14]

secondary importance as a downstream oxidation catalyst provides more than sufficient compensation for this drawback.

Optimum results are obtained when the start of injection is optimized across the entire map with respect to load and engine speed. This can be accomplished by using an electronic control system.

Chamber volume and chamber geometry: Swirl-chamber engines should be designed with a ratio of approx. 1.0 of the auxiliary chamber volume to the main chamber volume. *Figure 3.53* shows an example of how this ratio affects emissions.

This diagram underlines that a reduced swirl-chamber volume yields a slightly higher smoke number as well as an appreciable reduction of nitrogen oxide emissions across a wide range of the map. When the ratio of swirl chamber to main combustion chamber volume approaches unity, the smoke number increases sharply and NO_x emissions increase again at the same time. On the basis of the above findings, the ratio of the swirl chamber volume to the main combustion chamber volume of real engines is therefore set at 1.07 to 1.15.

In addition to the volume ratio, swirl chamber geometry is an important factor governing NO_x emissions. The heating curve can be modified by varying diameter, height and transfer cross-sectional area. The heating curve represents the energy quality released per degree of crank angle. If energy conversion proceeds with a certain delay and reaches its maximum only at a late point, NO_x emissions are reduced (*Fig. 3.54*). This, however, is accomplished at the expense of poorer efficiency.

To ensure good mixing, the shape and volume of the swirl chamber have to be optimized for the maximum possible swirl energy [3.32]. As in the case of precombustion engines, the position and geometry of the glow plug are important parameters. High swirl energy levels are obtained by using a thin glow plug that is recessed to a relatively large extent to disturb the flow conditions as little as possible. *Figure 3.55* shows the effects of projection depth and glow plug diameter on fuel consumption and smoke number [3.32].

b) Precombustion engine: The optimum pre-chamber volume relative to the main combustion chamber volume of precombustion engines is smaller than that of swirl-chamber engines (approx. 35 to 40%). If the pre-chamber is too small, the oxygen contents are no longer sufficient to ensure smoke-free combustion. HC, CO and particulate emissions will then increase as well. Due to the fact, however, that both pressure increase and peak temperature are reduced, NO_x emissions are lower.

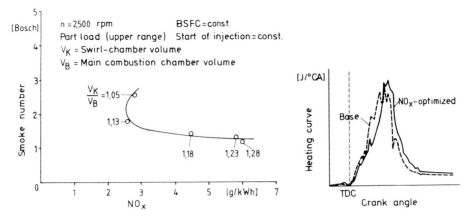

Fig. 3.53 Effect of chamber volume on emission characteristics [3.9]

Fig. 3.54 Heating curve of swirl chamber with optimized NO_x level [3.31]

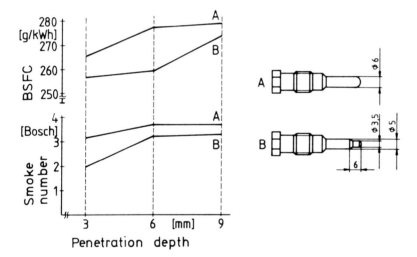

Fig. 3.55 Effect of glow plug configuration on fuel consumption and smoke number [3.32]

The geometrical design of the transfer passage is of particular importance. Excessive diameters mean that gas speeds are lower, i.e. the kinetic energy may not be sufficient to provide satisfactory mixture distribution and preparation in the main combustion chamber. If diameters are too small, they may become restricted due to clogging, and too much time may be required to transfer the charge from the auxiliary chamber into the main combustion chamber.

Other important design features are the auxiliary chamber position relative to the main chamber as well as position and shape of the glow plug and the injection jet angle. Best results are obtained if the auxiliary chamber is in a central position with regard to the main chamber.

One example of this approach is a design that features inclined injection instead of injecting the fuel centrally into the pre-chamber (*Fig. 3.56*) and that produces significant improvements in exhaust characteristics [3.11]. The start of injection is similar to that of swirl-chamber engines.

3.2 Diesel engines

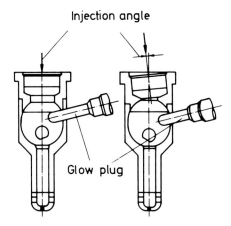

Fig. 3.56 Effect of engine-related measures on NO$_x$ and particulate emissions [3.11]

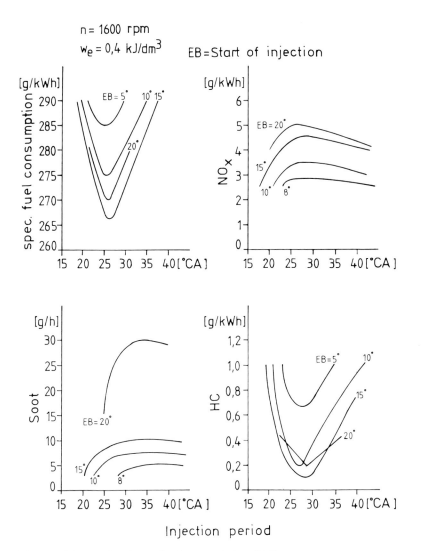

Fig. 3.57 Injection duration and pollutant emissions [3.3]

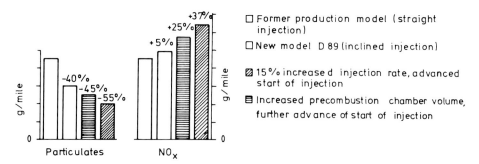

Fig. 3.58 NO$_x$ and particulate emissions measured in CVS tests [3.11]

Injection duration is another important parameter. Pump, pipe and injection nozzle should ensure that the fuel quantity is injected within a specified span of time. A characteristic feature of the injection duration is that there is no setting that will allow all pollutant components as well as fuel consumption to be minimized at the same time. To underline this point, *Fig. 3.57* shows an example of a supercharged engine.

Figure 3.58 shows the effect of the above measures on NO$_x$ and particulate emissions of a pre-chamber engine as determined by a CVS test. Improved particulate emissions are contrasted by impaired NO$_x$ emission figures.

3.2.1.2 Direct injection diesel engines

In the case of direct-injection diesel engines, a basic distinction is made between air-dispersion and wall-film dispersion processes. The air-dispersion process primarily distributes the fuel in the air whereas the wall-film dispersion process disperses the fuel on the cylinder wall. The fuel is then detached from the wall under the action of air swirl and is mixed with the combustion air.

Selection of the injection process has a direct effect on emission characteristics and fuel consumption. Air-dispersion processes offer advantages in terms of fuel consumption, HC emissions, smoke number and, hence, power density. Wall-film storage or wall-film dispersion processes only offer better NO$_x$ emissions and reduced combustion noise due to poorer fuel utilization (slower combustion at lower combustion temperatures). Air-dispersion processes are usually considered a better compromise and will therefore be described in more detail in the below section.

The curves of NO$_x$ and HC emissions, smoke number (SN) and fuel consumption as a function of specific work are shown for both processes used in a direct-injection diesel engine in *Fig. 3.59*.

As explained above, air-dispersion processes offer benefits with regard to HC emissions. In addition to achieving lower smoke numbers, this also has a positive effect on fuel economy, particularly in the upper load range.

a) Swirl inlet ports: Tuning of swirl, injection hydraulics and combustion chamber configuration are essential to a successful air-dispersion combustion process. The swirl shape also determines whether air and fuel can be mixed successfully. Each combustion chamber type, i.e. each type of piston bowl, is characterized by a specific optimum swirl rate (e.g. in terms of the smoke number [3.6]) that is also dependent on engine speed. Above and below this degree of swirl, a sharp increase of the smoke number has to be reckoned with. *Figure*

3.2 Diesel engines

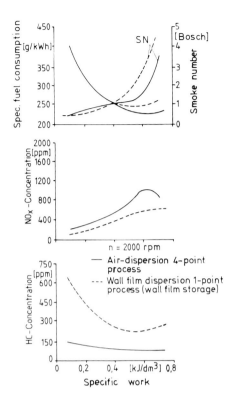

Fig. 3.59 Emission characteristics of air and wall-film dispersion (wall-film storage) processes [3.10]

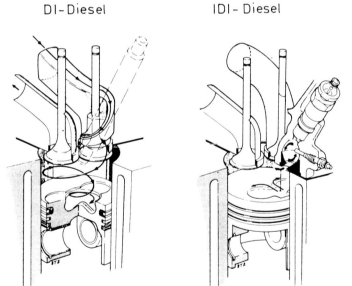

Fig. 3.60 Swirl port and air flow ducting for different diesel combustion processes [3.26]

3.60 shows the basic design of a swirl port and the air ducting into the piston bowl that is dictated by this port shape.

Figure 3.61 shows the effect of swirl on the emission level of the major pollutant components. This diagram shows the results obtained with a single-cylinder engine featuring a variable swirl port [3.34].

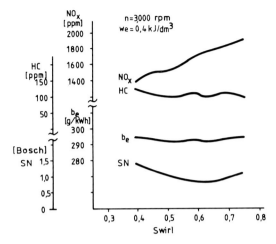

Fig. 3.61 Effect of swirl on exhaust emissions and fuel consumption [3.16]

To keep NO_x emissions low, swirl must be reduced. The effect of swirl on fuel consumption and HC emissions is not very pronounced. Furthermore, swirl also affects engine power. The higher the swirl, the lower the output since the cylinder charge is reduced.

b) Bowl shape and bowl location: In contrast with modern spark-ignition engine designs, the bowl is located in the piston. *Figure 3.62* shows some combustion chamber bowl configurations. Bowl characteristics are particularly favorable if the bowl is rotationally symmetrical to the piston axis since this allows swirl disturbances to be avoided. For reasons of geometry, however, this is only possible on multi-valve engines as this design requires the injection nozzle to be located centrally. A bowl restricted along its edges (*Fig. 3.62*) offers benefits with regard to NO_x emissions. A raised section in the center also promotes favorable emission characteristics.

Other design-related effects on NO_x and HC emissions are dictated by the position of the uppermost piston ring, i.e. the height of the top piston land and the resulting dead volume.

c) Current state of the art and potential for further development: In 1989, the first turbocharged diesel engine with direct injection designed to meet EEC emission regulation 88/436-EEC (Annex 25) was introduced on a vehicle with a mass classification of 3625 lbs, a swept volume of 2.5 l and an output of 88 kW [3.16] [3.17] [3.18] [3.19].

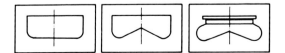

Fig. 3.62 Types of piston bowls

3.2 Diesel engines

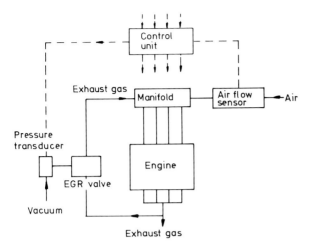

Fig. 3.63 Circuit diagram of an exhaust recirculation system [3.20]

Further development was since undertaken on this engine that now meets the stringent emission limits to Annex 23 [3.20]. The original design of this engine already featured intercooling and electronic diesel injection control of fuel quantity and start of ignition but was improved further by exhaust gas recirculation controlled by an air flow meter designed to reduce NO_x emissions.

A diesel oxidation catalyst was implemented to reduce HC, CO and particulate emissions.

The exhaust recirculation rate is specified for each map point and is controlled to avoid excessive HC, CO and, in particular, soot emissions that would be caused by excessive air deficiency. *Figure 3.63* shows the control circuit of the exhaust recirculation system.

Depending on the map point, this system allows recirculation rates of up to 50% to be achieved. The particulate emissions without EGR are so low on direct-injection diesel engines that even with EGR, no unduly high emission levels are produced. *Figure 3.64* shows the effect of exhaust recirculation on smoke number, fuel consumption and emissions. The design strategy is highlighted by the engine speed operating point) (*Fig. 3.65*). The unusable range in this figure is due to an insufficient difference between intake and exhaust pressure.

The diesel oxidation catalyst used here is characterized by a small pressure differential to avoid any significant effect on engine output. *Figure 3.66* shows the reduction of HC and CO emissions for two catalytic converter versions with different coatings. From exhaust temperatures as low as 200 °C, version B achieves a conversion efficiency of approx. 40% for both pollutant components.

An important coating parameter is the reduction of sulphate formation. Coatings that allow only very limited SO_3 formation have since been introduced (e.g. coatings based on palladium). CVS tests have yielded the following conversion efficiencies for this concept:

- HC: up to 40%
- CO: up to 50%
- Particulates: up to 10%.

As far as particulates are concerned, polycyclic aromatic hydrocarbons (PAK) attached to soot are reduced to a particularly great extent. The low HC emissions that remain when the engine is at operating temperature (25% or below) produce far less exhaust odor since aldehyes are reduced at the same time.

NO_x is not reduced by the oxidation catalyst in an atmosphere that is rich in oxygen and low in hydrocarbons. If this type of catalytic converter is used, however, conversion of HC

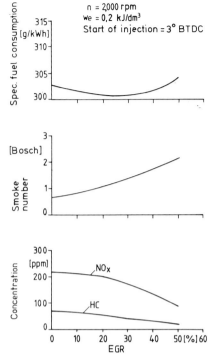

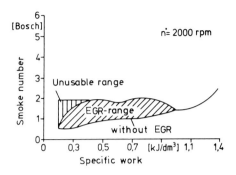

Fig. 3.65 Design strategy of exhaust recirculation [3.20]

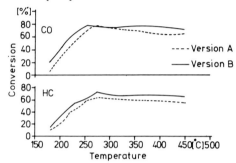

Fig. 3.64 Effect of exhaust recirculation on exhaust emissions and fuel consumption [3.20]

Fig. 3.66 Conversion efficiencies of a production oxidation catalyst used on a diesel engine [3.20]

emissions increases the degree of freedom available for optimizing the engine since an extremely retarded start of injection could otherwise only be used at the expense of a critical increase of HC emissions (*Fig. 3.69*).

The effect of the start of injection on NO_x levels is obvious: The more injection is retarded, the lower are NO_x emissions. As we approach TDC, the gradient of NO_x concentrations levels off, and if the start of injection is shifted further past TDC, HC emissions and fuel consumption increase steeply.

The smoke number increases slightly from the advanced start of injection to just before TDC and then drops off again (cf *Fig. 3.67*).

The optimum start of ignition at this map point, taking production tolerances into account, is 1 deg. crank BTDC. This approach to determine the optimum start of injection was put into practice on selected map points. *Figure 3.68* shows the resulting start of injection map.

Figure 3.69 shows the advanced state of development of diesel engines with oxidation catalysts. A CVS test that was commenced under cold starting conditions at $-7\,°C$ has shown that HC emissions were only insignificantly higher than at $22\,°C$. As is to be expected, NO_x figures improve slightly. CO is impaired to a greater extent but still remains far below the admissible limit [3.20].

The specific fuel consumption figures obtained are shown in the map in *Fig. 3.70*. Fuel economy of this engine developed in accordance with Annex 23 is only slightly poorer than that of an engine designed to comply with Annex 25. This is partly due to the later start of injection that was required to keep NO_x emissions down. Yet this engine shows the best fuel economy by far that has so far been achieved on an engine designed according to Annex 23.

3.2 Diesel engines

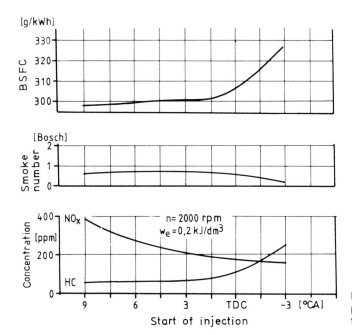

Fig. 3.67 Effect of start of injection on pollutant emissions and fuel consumption [3.20]

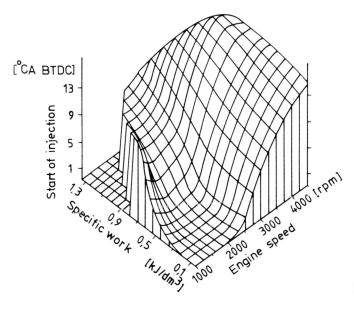

Fig. 3.68 Start of injection map [3.20]

Multi-valve technology with centralized location of the injection nozzle and piston bowl contributes to a further reduction of fuel consumption.

Figure 3.71 shows the emissions of the individual pollutant components in a CVS test, with emissions remaining below the limit with sufficient certainty. Since early 1992, emissions have also been lower than the limits specified by German particulate emission regulations (0.08 g/km).

CO_2 emissions of direct-injection diesel engines should also be noted. They are approx. 15% below those of a pre-chamber engine, and approx. 30% below those of a modern

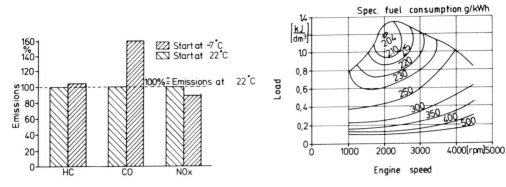

Fig. 3.69 Comparison of emissions under cold-starting and ambient temperature conditions [3.20]

Fig. 3.70 Fuel consumption map [3.20]

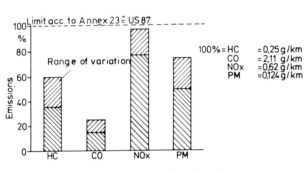

Fig. 3.71 Pollutant emissions in CVS test [3.20]

Fig. 3.72 Comparison of CO_2 emissions of different engine designs [3.20]

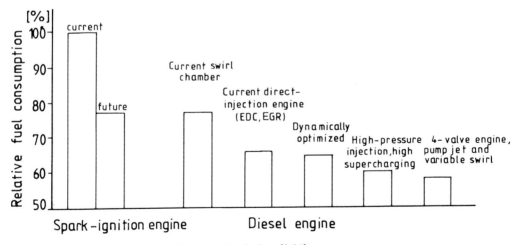

Fig. 3.73 Fuel savings potential of different engine designs [3.21]

spark-ignition engine (cf. *Fig. 3.72*). This feature also contributes to reducing the greenhouse effect.

Figure 3.73 shows the fuel savings potential of a diesel engine compared to a spark-ignition engine according to ECE cycle testing [3.21]. Both engine versions show a significant potential for cutting down fuel consumption.

3.2 Diesel engines

Emission-control measures	Specific fuel consumption	Pollutant emissions NO	HC	Soot	Part.
Start of injection (retard)	↑	↓	↑	↑	↑
Exhaust gas recirculation, hot	—	↓	↓	↑	↑
Exhaust gas recirculation, cooled	—	↓	—	↑	↑
Supercharging	—	↑	↓	↓	↓
Intercooling	↓	↓	—	↓	↓
Particulate filter	↑	—	—	↓	↓
Diesel oxidation catalytic converter	—	—	↓	—	↓

Fig. 3.74 Parameters affecting pollutant emissions [3.63]

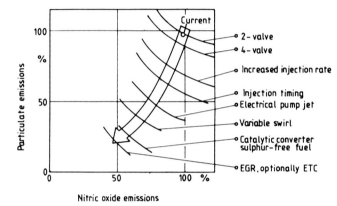

Fig. 3.75 Potential for emission reduction of direct-injection diesel engines [3.21]

For reasons of fuel economy, however, modern diesel engines should be designed as direct-injection engines with supercharging and intercooler.

This will allow good CO and HC exhaust levels to be obtained whereas NO_x and particulate emissions will have to be reduced yet further to comply with more stringent emission standards.

Figures 3.74 and *3.75* [3.21] [3.63] show ways to reduce emissions on direct-injection diesel engines, with the implementation sequence of these measures being variable.

All the measures described improve operational characteristics considerably. The development costs involved are rather high, though. As is the case with spark-ignition engines, future DI diesel engines will be high-tech engines with an accordingly high price tag.

Figure 3.76 shows a strategy to reduce NO_x particulates to meet the stringent US emission limits applicable from 1994 ($NO_x \leq 0.4$ gpm and particulates ≤ 0.08 gpm, also refer to Chapter 8.4.1.1).

Starting with vehicle emission figures achievable with EGR but without exhaust aftertreatment (upper curve, *Fig. 3.76*), the use of either an oxidation catalyst or a particulate trap will allow emissions to be improved. The example shown indicates that the lowest particulate levels can be achieved down to a NO_x level of approx. 0.3 to 0.4 gpm if an oxidation catalyst is fitted. To achieve yet lower NO_x emissions, a particulate trap and exhaust gas recirculation should be used.

It has been found, however, that irrespective of the exhaust aftertreatment strategy adopted, the targeted legal emission standards are not complied with. To achieve a develop-

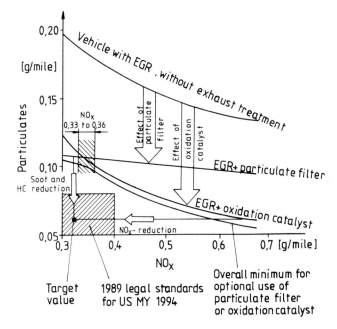

Fig. 3.76 Exhaust aftertreatment strategies used to reduce NO_x and particulate emissions [3.22]

ment target of 0.32 grams of NO_x/mile and 0.07 grams of particulates/mile, two basic development strategies may be envisaged:

- Use of a particulate trap and incorporation of additional measures that will reduce soot and HC emissions yet further,
- Use of an oxidation catalyst and additional measures for further NO_x reduction.

If space requirements, high cost, loss of engine power and regeneration problems of particulate traps are taken into account, the latter approach of using an oxidation catalyst appears to be preferable.

From today's point of view, it should be possible to achieve further reductions of NO_x emissions as shown in *Fig. 3.77*.

The measures required for this purpose are:

- A Denox catalyst that offers NO_x reductions in excess of 50% even at the current development stage (cf. Chapter 5),
- Improved mixture formation using four-valve technology [3.68]. The centralized bowl position allows variable swirl to be achieved (one swirl port, one filling port) and provides a central injection nozzle position, *Fig. 3.78* [3.37] [3.59],
- Injection pressure $> 1,500$ bar, e.g. by using pump-nozzle combinations or common rail systems in conjunction with limited swirl,
- Optimized mixture formation achieved by controlling the injection rate control (*Fig. 3.79*). This diagram shows the required injection quantity as a function of time, with time being composed of the pre-injection and the main injection phases. If the injection quantity is determined for each map point and made available for driving across a suitable electronic system, NO_x reductions of up to 30% are achieved,

3.2 Diesel engines

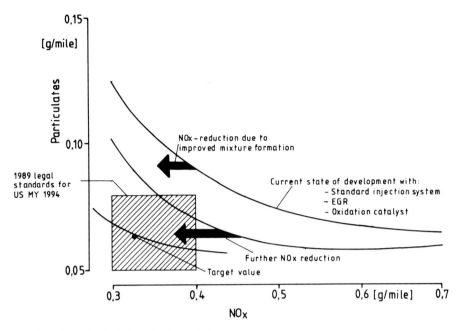

Fig. 3.77 Potential for NO_x reduction [3.22]

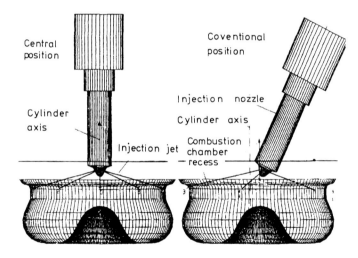

Fig. 3.78 Position and injection jet of injection nozzle [3.37]

- Increasing the charging pressure (supercharging),
- Use of variable turbine geometry of exhaust turbocharger,
- Mass flow control in the part-load range,
- Use of low-sulphur fuel.

Solutions for some of the above approaches (e.g. injection rate control) are not yet available for production vehicles. Further "exotic" measures to reduce NO_x emissions are:

- Injection of water and water-fuel emulsions, with injection of water-fuel emulsion causing NO_x emissions to be reduced by approx. twice the factor achievable with water injection [3.57]. The effect of injecting water along with the charge air or intake air also is about twice as poor as the effect of injecting water-fuel emulsions [3.58]. A water content

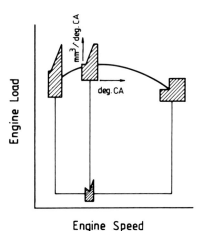

Fig. 3.79 Mixture formation with injection rate control [3.22]

of approx. 25% relative to the fuel mass flow allows NO_x emissions to be reduced by almost 50%.
- Intercooling of the recirculated exhaust, thus allowing NO_x emissions to be reduced by 50%.

Recent injection nozzle developments tended to favor full-load/part-load nozzles. With this setup (with "two" nozzles in one nozzle), nitrogen oxide emissions may be reduced below 0.4 gpm in FTP-75.

3.2.2 Compression ratio

The compression ratio ε of passenger vehicle diesel engines is approx. $\varepsilon = 20$ to 23. Swirl-chamber or pre-chamber engines usually have slightly higher compression ratios than direct-injection engines. Increasing the compression ratio also helps to improve HC and particulate emissions, especially since the content of soluble particulates at part throttle is reduced by this process. On the other hand, this also increases friction work and, hence, fuel consumption and NO_x emissions. The limits are imposed by soot emissions at full throttle.

Figure 3.80 shows the effect of the compression ratio on HC emissions vs start of injection at one particular part-load point.

3.2.3 Fuel injection hydraulics

The hydraulics of fuel injection is another essential parameter having an impact on pollutant formation in the diesel engine. Direct-injection diesel engines, in particular, benefit from high injection pressures with regard to fuel consumption and pollutant emissions. Different injection systems such as common rail injection systems, Junit injectors, pump-line-nozzle systems etc. are available for this purpose.

The injection nozzle geometry also is an essential feature in conjunction with injection hydraulics. Direct-injection engines use either single-hole or multi-hole nozzles. HC emissions can be improved by reducing the pollution volume (sac hole volume) of the injection nozzle as this limits dripping of the nozzle.

Particular advantages are offered in this context by the valve cover orifice nozzle shown in *Fig. 3.81* along with other nozzle configurations.

3.2 Diesel engines

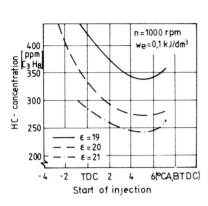

Fig. 3.80 HC emissions of a direct-injection diesel engine

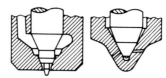

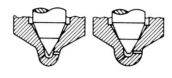

Fig. 3.81 Injection nozzle configurations (sac hole and valve cover orifice nozzles)

The number of injection holes has a considerable influence on smoke formation and on NO_x emissions. At an identical smoke number, NO_x emissions are lower if, for example, a 4-hole nozzle is replaced by a 5 or 6-hole nozzle. *Figure 3.82* highlights this relationship. It must be borne in mind, however, that acoustics (combustion noise) are impaired when the number of holes is increased since this entails an increased ignition delay and, hence, a higher initial energy conversion rate.

Figure 3.82 also shows the effect of swirl on the smoke number and on NO_x emissions, respectively. In this case, swirl is defined as the ratio of circumferential velocity to axial velocity.

Independent of the number of injection holes of the nozzle, a higher swirl number yields lower smoke numbers and NO_x emissions.

The effect of achieving improved pollutant emission figures by increasing swirl is largely dependent on the actual injection pessure level. Pre-injection has a positive effect on acoustics and on emission characteristics. A twin-spring nozzle holder was used to build a pre-injection system. A very small initial stroke injects a small quantity of fuel; the main quantity is injected during the second, longer stroke.

The basic design of this type of twin-spring nozzle holder used for a 2.5-liter turbocharged engine with direct fuel injection is shown in *Fig. 3.83*.

The effectiveness of this multi-stage injection process is based on the fact that a small fuel quantity is made to react initially with a high excess air rate. This improves mixture

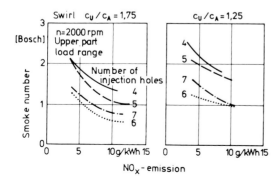

Fig. 3.82 Effect of swirl and number of injection holes on emission characteristics [3.9]

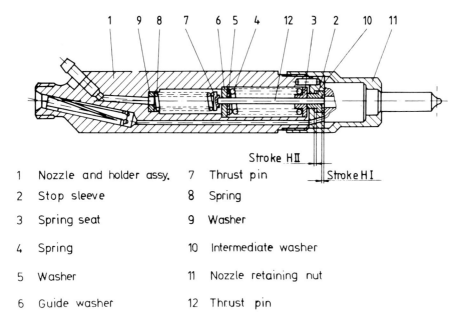

Fig. 3.83 Twin-spring nozzle holder for a diesel engine with direct injection [3.18]

1 Nozzle and holder assy.	7 Thrust pin
2 Stop sleeve	8 Spring
3 Spring seat	9 Washer
4 Spring	10 Intermediate washer
5 Washer	11 Nozzle retaining nut
6 Guide washer	12 Thrust pin

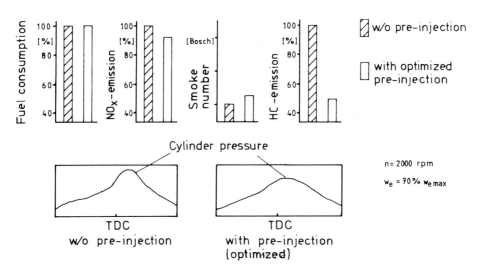

Fig. 3.84 Pre-injection in a direct-injection diesel engine [3.9]

formation and reaction conditions of the subsequently injected fuel quantity considerably. As a result, HC emissions decrease sharply, and nitrogen oxide emissions are also improved.

Figure 3.84 shows the relative change achieved with pre-injection designs in terms of the pollutant components affected. Pre-injection produces a more steady pressure flow inside the cylinder, and the pressure gradient is reduced at the same time. Since the ignition lag of the main injection quantity is reduced, combustion noise level is lowered as an additional benefit.

The injection pressure is, along with the injection nozzle geometry, another efficient means of reducing soot and nitrogen oxide emissions by a significant margin.

3.2 Diesel engines

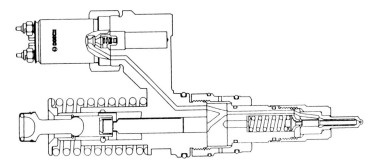

Fig. 3.85 Pump-nozzle unit [3.37]

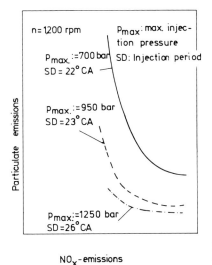

Fig. 3.86 Effect of injection pressure on particulate and NO_x emissions [3.79]

Higher injection pressures combined with smaller injection holes improve atomization and therefore have a positive effect on mixture preparation. High injection pressures may be achieved e.g. by using a pump-nozzle element. The design of an element of this type is shown in *Fig. 3.85* [3.37]. Systems of this type are suitable for injection pressures in excess of 1,500 bar. High injection pressures can also be achieved by using common rail injection systems.

The higher the injection pressure, the lower are nitrogen oxide and particulate emissions (*Fig. 3.86*). The injection pressures of approx. 900 bar achievable today with DI diesel engines at a max. engine speed of approx. 4,700 rpm offer further potentials for emission reduction. The optimum injection pressures in terms of pollutant emissions and fuel consumption are far above 1,500 bar [3.37].

3.2.4 Electronic control of diesels

The wide range of ever-increasing requirements imposed, among other reasons, with the aim of reducing emissions, suggests the use of electronic fuel injection control systems in diesel engines in the same way they are used for spark-ignition engines. Distributor-type injection pumps are widely used for this purpose.

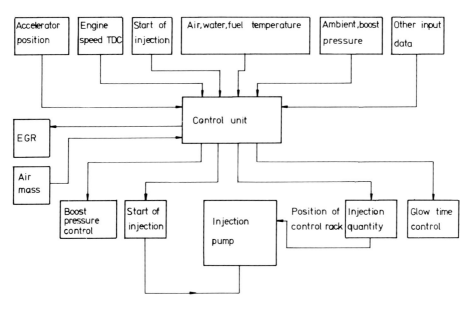

Fig. 3.87 Block diagram of diesel control system [3.18]

Figure 3.87 shows the block diagram of an electronic diesel control system. This type of control is based on detecting specific operational parameters using suitable sensors. The major parameters are:

- Information on engine load condition, i.e. accelerator position, e.g. by using an electronic accelerator (dispensing with mechanical linkage between accelerator and distributor-type pump),
- Engine speed, engine temperature,
- Start of injection,
- Load and ambient pressure of supercharged engines.

The above parameters are used as inputs for the control system. The control unit supplies output signals that are dependent on the above values and on the maps or curves stored. These, in turn, are coupled with actuators to modify parameters such as:

- Fuel quantity,
- Start of injection,
- Glow pencil control,
- EGR (control of this parameter is performed depending on load and engine speed).

Close matching of the start of injection to all operational conditions of the engine is an important requirement to ensure optimum mixture formation and an optimum combustion process. Due to the fact that NO_x, HC, particulate and soot emissions are heavily dependent on the start of injection, this parameter must meet the specifications closely in order to achieve low emissions. The more the injection start is retarded, the lower are the NO_x emissions in the relevant range. Hydrocarbon and particulate emissions, however, react in reverse. *Figure 3.88* shows the effect of injection start on HC and particulates and the different scatterbands of electronic and mechanical control systems.

It becomes evident that emission characteristics are greatly improved if the scatterband is narrowed down. Good reproducibility of the start of injection is ensured by a closed-loop control circuit. The actual start of injection measured with a needle movement sensor at the injection nozzle is compared with a temperature-dependent map of the target

3.2 Diesel engines 69

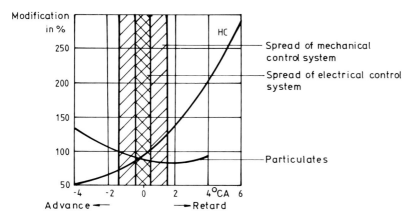

Fig. 3.88 Effect of scatterband of diesel control system on particulate and HC emissions of a pre-chamber engine [3.65] [3.77]

injection start and is corrected accordingly. This allows accuracies of approx. 2 deg. crank to be achieved. The second important input is performed by electronic volume control. The control unit includes the map data required for this purpose. The map data are then compared with the actual values determined via sensors, allowing the fuel quantity to be adjusted accordingly. The quantity actually released is limited by the smoke map. Tuning is performed in such a manner that the pre-set smoke limit will not be exceeded at any operating point.

3.2.5 Exhaust recirculation

Pollutant emissions of diesel engines can also be reduced if exhaust recirculation is adopted for the part-load range, and recirculation rates may in this case be far higher than with spark-ignition engines (also refer to 3.1.1.4). Cooling the recirculated exhaust gas allows NO_x emissions to be reduced even further.

Direct-injection diesel engines achieve recirculation rates of more than 60%, and pre-chamber engines reach figures of up to 30%. Control should focus on nitrogen oxides as they can be reduced to a particularly great extent in the part-load range. To a lesser degree, this also applies to HC emissions. Depending on the recirculation rate in the higher-load range, particulate emissions, on the other hand, may increase due to the reduced oxygen supply available during combustion.

The absolute figures achieved on diesel engines with direct injection, however, are relatively low. *Figure 3.89* shows the levels with respect to specific work.

Figure 3.90 compares emissions of pre-chamber engines and direct-injection engines with and without exhaust recirculation. With regard to the NO_x particulate gap (interdependence between NO_x and particulate emissions), the direct-injection diesel engine without EGR fares rather worse than a pre-chamber engine. Since direct-injection diesel engines are better compatible with EGR systems, however, this disadvantage can be compensated.

3.2.6 Diesel engine potential in terms of emissions and fuel consumption

Both pre-chamber engines and, in particular, direct-injection diesel engines offer a significant potential for reducing fuel consumption and exhaust emissions. Multi-valve engines, for example, will yield improved mixture preparation as the injection nozzle can be located

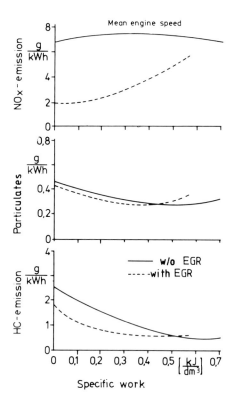

Fig. 3.89 Basic effect of exhaust recirculation on pollutant emissions

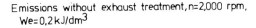

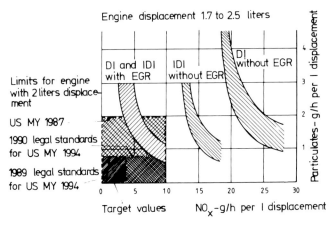

Fig. 3.90 Potential use of exhaust recirculation systems for various diesel engine designs [3.22] [3.77]

in a suitable position and the inlet ports can be used accordingly. Swirl-chamber engines using three valves per cylinder are already in large-scale production.

4-valve precombustion engines were introduced in 1993. Direct-injection diesel engines with 4-valve technology are currently being developed. The future potential use of lean-burn catalytic converters and particulate traps open up new strategies for engine design, e.g. with regard to the NO_x/particulate gap, and therefore offer new degrees of freedom in engine tuning.

Increasing the injection pressures of direct-injection diesel engines (pump-nozzle design, common rail system), injection rate control, cooled EGR, variable swirl and

3.3 Characteristic engine emission maps

optimized cylinder supercharging are some additional elements that allow both fuel consumption and emissions to be reduced.

The introduction of modified fuels will be an essential step towards this goal, either by reducing the content of aromatic substances, raising the cetane rating or by reducing the sulphur content drastically (this latter measure being an important step towards reducing particulate emissions).

Future diesel engines will certainly be direct-injection engines and will be designed as high-technology engines.

The fuel economy benefits will, in terms of CO_2 emissions, be another bonus contributing to increasingly widespread use of diesels in the future. Similar to spark-ignition engines, the overall potential for reducing fuel consumption amounts to more than 30%, and, in the same way as with spark-ignition engines, emissions will remain below future emission standards. Again, the effect of these measures will be enhanced if extensive measures to reduce fuel consumption are implemented on the vehicle itself.

3.3 Characteristic engine emission maps

The following illustrations show the emission characteristics of spark-ignition and diesel engines. *Figures 3.91* and *3.92* show the characteristic map of HC and NO_x emissions of a spark-ignition engine. The HC emissions measured on this four-valve engine are relatively low for a production engine. The NO_x values were determined at the same engine setting. *Figures 3.93, 3.94* and *3.95* show the CO, HC and NO_x values of another four-valve spark-ignition engine that may be representative of present-day engines.

Figures 3.96 to *3.103* show characteristic maps of major pollutant emissions of diesel engines using different combustion processes. A supercharged three-valve swirl-chamber engine is compared with a direct-injection engine.

Figures 3.104 to *3.106* show the major emissions, i.e. HC, HO and NO_x, in terms of specific work for SI, swirl-chamber diesel and DI diesel engines. The DI diesel is fitted with exhaust turbocharger and intercooler. The scatterband represents the engine speed range from 2,000 to 6,000 rpm for spark-ignition engines and from 2,000 to 4,5000 rpm for diesel engines. It becomes clear that the spark-ignition engine has the highest raw emissions.

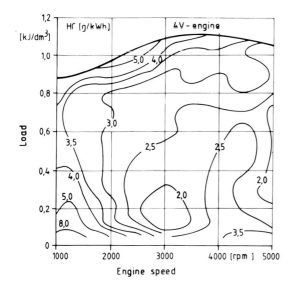

Fig. 3.91 HC emissions of a four-valve engine

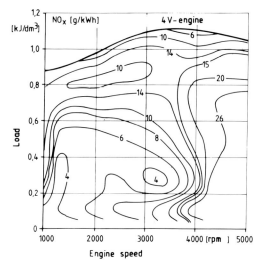

Fig. 3.92 NO$_x$ emissions of a four-valve engine

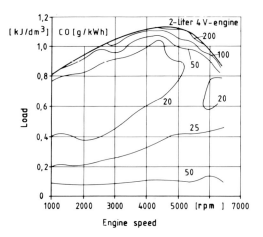

Fig. 3.93 CO emissions of a representative four-valve spark-ignition engine

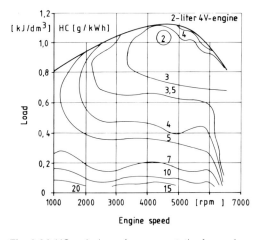

Fig. 3.94 HC emissions of a representative four-valve spark-ignition engine

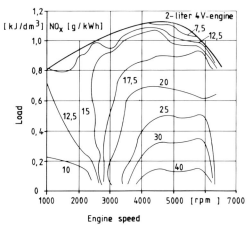

Fig. 3.95 NO$_x$ emissions of a representative four-valve spark-ignition engine

When comparing the different diesel engine designs, the raw emissions of the swirl-chamber engine (with the exception of CO) are lower. The sharp increase of CO emissions of spark-ignition engines at high specific work rates is due to full-throttle enrichment that allows output to be increased by approx. 10%.

3.4 Long-term emission stability

Ageing, wear, changing tolerances, deviations from basic settings and defective components produce changes in the emission characteristics during the service life of the engine. This affects spark-ignition engines to a far greater degree than diesel engines. In this context, a distinction has to be made between emissions of in-use vehicles and type approval

3.4 Long-term emission stability

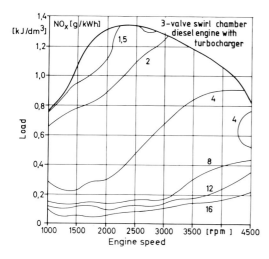

Fig. 3.96 NO$_x$ emissions of a swirl-chamber engine

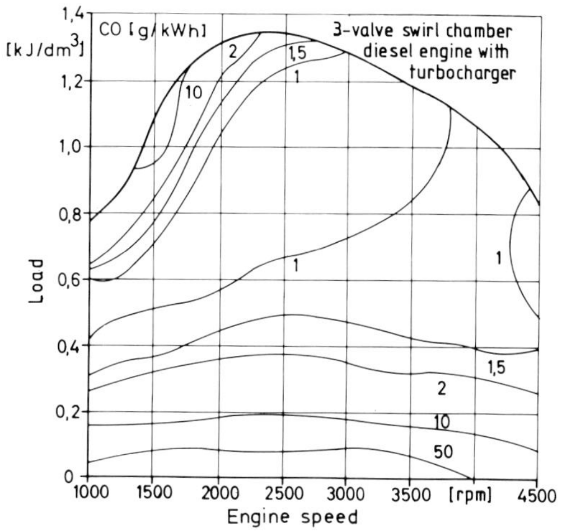

Fig. 3.97 CO emissions of a swirl-chamber engine

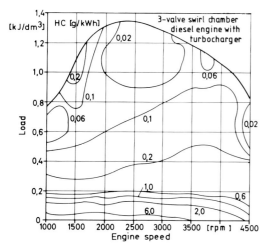

Fig. 3.98 HC emissions of a swirl-chamber engine

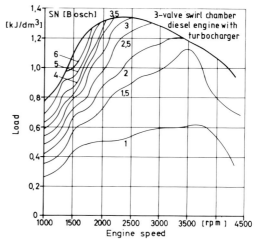

Fig. 3.99 Smoke numbers of a swirl-chamber engine

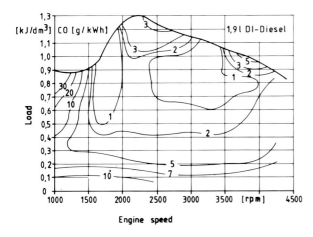

Fig. 3.100 CO emissions of a direct-injection diesel engine

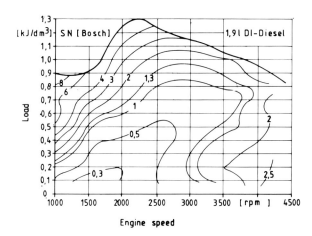

Fig. 3.101 Smoke numbers of a direct-injection diesel engine

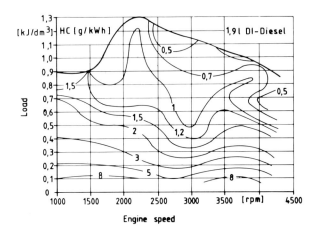

Fig. 3.102 HC emissions of a direct-injection diesel engine

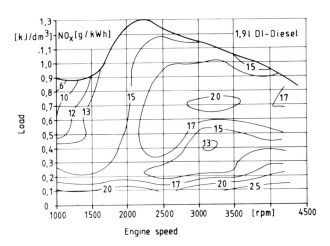

Fig. 3.103 NO_x emissions of a direct-injection diesel engine

3.4 Long-term emission stability

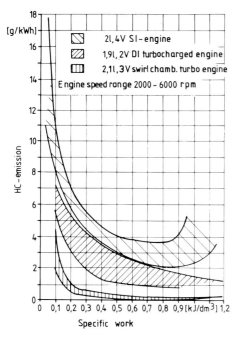

Fig. 3.104 HC scatterbands of diesel and spark-ignition engines

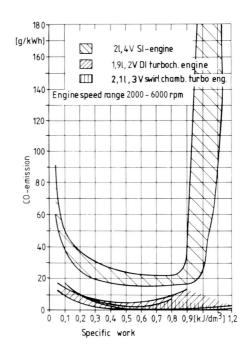

Fig. 3.105 CO scatterbands of diesel and spark-ignition engines

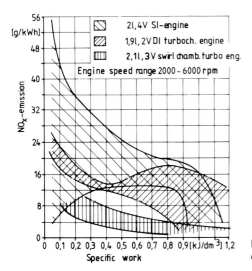

Fig. 3.106 NO_x scatterbands of diesel and spark-ignition engines

standards of new vehicles. The changes are caused mainly by ageing of the catalytic converter and λ sensor, chemical and thermal effects on these components, functional changes of mixture formation and ignition as well as a lack of maintenance etc.

The type approval test required for vehicles exported to the USA requires a long-term exhaust test (Chapter 8) that is used to prove emission consistency, taking corresponding deterioration factors (DF) into account. *Figure 3.107* shows the curves for major emissions of a diesel engine during such a long-term emission test [3.42].

The above effects occurring during real-life operation can only insufficiently be accounted for by this type of long-term test and are not reproduced by deterioration factors

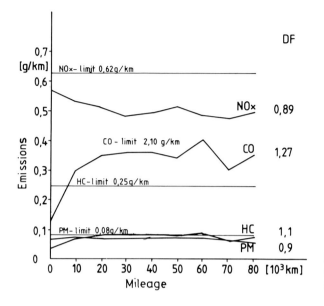

Fig. 3.107 Exhaust durability test run to determine deterioration factors of a diesel engine [3.42]

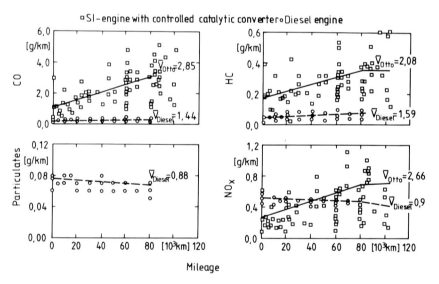

Fig. 3.108 Exhaust characteristics during engine life [3.60–3.74]

[3.43]. Deterioration of emission characteristics under real-life driving conditions was, among others, investigated by [3.44]. *Figure 3.108* shows the most important results.

As is evident from the diagrams, the scatterband of spark-ignition engines is far wider than that of diesel engines. As pointed out before, diesel engines also offer better long-term stability in their emission characteristics. Particulates and nitrogen oxides actually decrease during the life of the engine whereas CO and HC emissions increase to a far lesser extent than on spark-ignition engines. This is one detail which has so far been given only little attention when comparing spark-ignition and diesel engines.

4 Engine-related measures which reduce pollutant emissions and fuel consumption in two-strokes

Two-stroke engines have a very long tradition. Thanks to attractive features such as simplicity of design (no scavenging blower, no valve train), reduced space requirements, low weight (higher power density) and good vibrational comfort (double ignition frequency), they constituted a viable alternative to four-stroke engines for use in passenger cars (*Fig. 4.1*). In addition to coventional carburetor engines, an interesting design was e.g. the Goliath two-cylinder two-stroke engine with direct gasoline injection and loop scavenging (*Fig. 4.2*).

In the field of large-size engines (low-rpm uniflow-scavenged stationary diesel engines), two-stroke engines actually offer the highest effective efficiencies ever reached by any internal-combustion engine design. Efficiencies of more than 50% have been achieved on this type of engine. This corresponds to a specific fuel consumption of 160 to 170 g/kWh (*Fig. 4.3*).

Problems due to excessive fuel consumption and pollutant emissions that were caused, among other reasons, by scavenging losses and by the requirement to add oil to the fuel, have led to the elimination of two-stroke engines from passenger-vehicle applications.

In recent years, however, many passenger car manufacturers around the world have investigated new concepts to overcome the problems mentioned above and to exploit the benefits of this design, and it is therefore quite possible that the first two-stroke engines fitted with the required emission control systems and offering good fuel economy will enter production by the late 90s. Development work essentially focuses on the following main processes:

- Loop scavenging with port control, *Fig. 4.4*

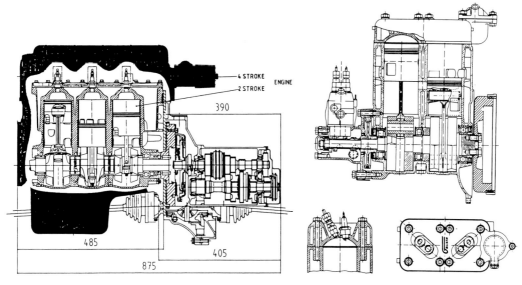

Fig. 4.1 Comparison of space requirements of two-stroke and four-stroke engines [Source: AVL]

Fig. 4.2 Goliath engine with fuel injection and loop scavenging [Source: AVL]

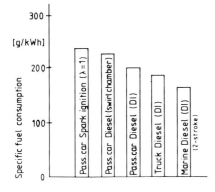

Fig. 4.3 Fuel consumption of internal combustion engines

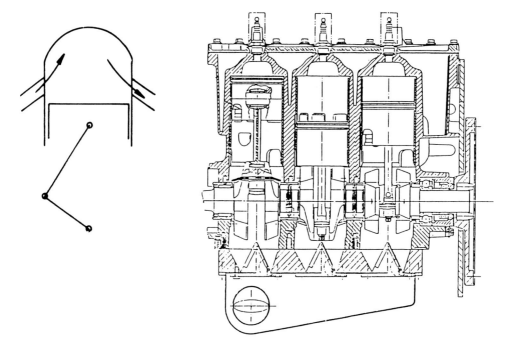

Fig. 4.4 Loop scavenging with port control [Source: AVL]

- Uniflow scavenging, *Fig. 4.5*
- Head scavenging, *Fig. 4.6*

Complexity, installed size and weight increase in the sequence listed above. Loop-scavenged engines have no valves, uniflow-scavenged engines use exhaust valves or opposed pistons with port control, and head scavenging designs are fitted with inlet and exhaust valves. In order to provide sufficiently sized charge changing areas, head scavenging designs preferably use two exhaust and two inlet valves.

Friction power and, hence, fuel consumption increase in the above order. Mechanical friction of loop-scavenged engines is very low as there is no valve train. In addition, charge changing work is far more limited than in engines with valve control (*Fig. 4.7*). This has a particularly positive effect on fuel efficiency at part throttle (*Fig. 4.8*).

If, on the other hand, a valve train is used as in the case of uniflow and head scavenging designs, part-load fuel consumption increases since the camshaft speed is twice as high as on four-stroke engines and friction power is higher as well.

4 Engine-related measures which reduce pollutant emissions and in two-strokes

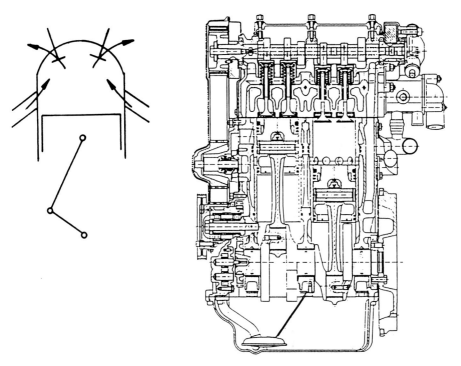

Fig. 4.5 Uniflow scavenging with valves and ports [Source: AVL]

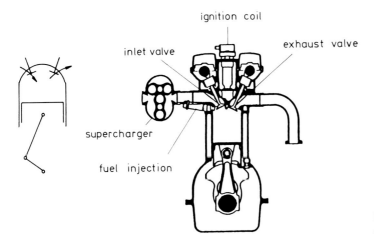

Fig. 4.6 Head scavening [Source: AVL]

Many designs use blower or head scavenging systems to keep engine oil out of the combustion chamber in an effort to reduce emissions and to increase power density (i.e. to increase cylinder charge).

The crankcase scavenging system that uses the area below of the working pistons, however, continues to play an important role thanks to its simplicity of design (also refer to orbital engine design, *Fig. 4.9*).

In order to achieve high power densities, compressors or screw-type compressors are often fitted. The above statements on the three cycles described apply both to two-stroke spark-ignition engines and to two-stroke diesel engines.

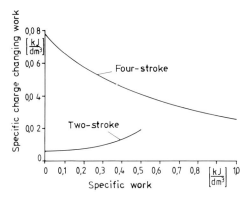

Fig. 4.7 Charge-changing work of two-stroke and four-stroke engines [Source: AVL]

Fig. 4.8 Part-load fuel consumption of two-stroke and four-stroke engines [Source: AVL]

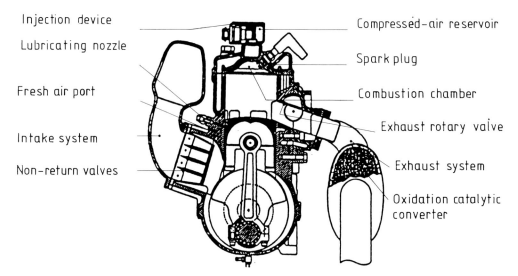

Fig. 4.9 Cross-sectional view of an orbital engine [Source: AVL]

4.1 Gasoline engines

Two-stroke gasoline engines offer an inherent advantage over four-stroke engines in terms of NO_x emissions. This is due to the operating cycle that involves a high residual gas rate and, therefore, lower combustion temperatures. The residual gas content may be as high as 70% in the lower part-throttle range and decreases to a level of approx. 20% as the specific work increases [4.3]. At full throttle, it is still significantly higher than the figures of four-stroke engines.

Operating the engine at a setting of $\lambda = 1$ as used with four-stroke engines is not possible due to the differences in the charge-changing process [4.4]. A three-way closed-loop catalytic converter therefore cannot be used in an efficient manner. Since lean-burn catalytic converters are currently not yet available, it is of particular importance to control and minimize NO_x raw emissions. Low NO_x emissions can be obtained at high (internal or external) EGR rates [4.5] and by exploiting lower load ranges [4.4]. One way of accomplishing this is to fit large-displacement engines to vehicles of low weight. Further

4.1 Gasoline engines

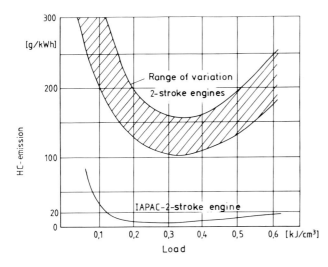

Fig. 4.10 HC emissions of a two-stroke engine

measures may include introducing suitable charge stratification processes, adapting the injection and charge changing strategies, and reducing pumping and frictional losses. All this contributes to low NO_x raw emissions [4.7] [4.8] [4.10].

HC emissions of two-stroke engines are essentially caused by scavenging losses. In this process, part of the fresh charge entering the cylinder short-circuits directly into the exhaust system. Since the exhaust gas temperature of two-stroke engines is higher than that of four-stroke engines, emission control benefits can be achieved e.g. by incorporating an oxidation catalyst.

Extremely low HC emissions can be reached e.g. by using suitable injection systems, inducing a controlled charge movement, charge stratification etc. [4.4]. The improvements obtained by these measures are indicated in *Fig. 4.10*.

For emissions and fuel economy reasons, only internal mixture formation should be considered for two-stroke engines. Fuel or a rich air-fuel mixture must be injected directly into the combustion chamber (high-pressure injection) after the charge-changing process has been completed. This helps to provide quality management similar to the process in four-stroke diesel engines, resulting in optimum combustion efficiency and, hence, in minimum fuel consumption. The resulting lean-burn operation has certain drawbacks in terms of emission characteristics, however, when compared to a four-stroke engine operated at $\lambda = 1$.

Mixture preparation across the entire load and rpm range is the main area of concern. Since the duration of the engine processes is far shorter than in the case of engines with external mixture formation, auxiliary measures to aid mixture preparation are not only beneficial to fuel injection but in fact such measures are essential. Two promising processes are currently under development:

- Air-assisted fuel injection for preparation of the fuel jet (*Fig. 4.9*),
- Mixture injection (*Fig. 4.11*).

The example of mixture injection will be used to explain the mixture preparation process in greater detail [4.1]. In the initial stage, fuel is injected into the closed mixing chamber and is premixed with hot air (*Fig. 4.11*).

This leaves more time for fuel preparation (injection – vaporization – combustion) than in direct-injection systems since fuel is prevaporized and can be introduced in a partially gaseous state (*Fig. 4.12*).

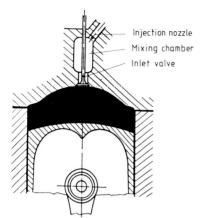

Fig. 4.11 Two-stroke engine with mixture injection [Source: AVL]

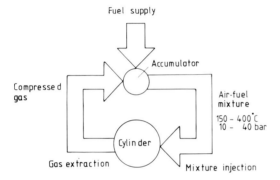

Fig. 4.12 Simplified diagram of mixture injection [Source: AVL]

This type of external fuel injection requires injection pressures of approx. 5 to 6 bar (up to 20 bar are actually possible). *Figure 4.13* shows the different strategies used for full-load and part-load injection processes.

Figure 4.14 summarizes the differences between air-assisted mixture injection and direct injection vs engine speed in terms of fuel consumption, hydrocarbon and NO_x emissions of a specific engine. It is evident that the benefits of mixture injection are particularly high at low engine speeds and low loads [4.10].

4.2 Diesel engines

In order to keep fuel consumption low, two-stroke diesel engines should be designed as direct-injection engines, i.e. pre-chamber designs should be avoided. *Figure 4.15* shows a cross-sectional view of a loop-scavenged engine, whereas *Fig. 4.16* shows an engine with uniflow scavenging. *Figure 4.17* represents a four-valve cylinder head of a head-scavenged engine.

Basically, the combustion process of a two-stroke diesel engine is subject to the same constraints that apply to four-stroke diesel engines [4.9] [4.11].

NO_x raw emissions are somewhat lower thanks to reduced peak combustion temperatures and low mean effective pressures. If an exhaust recirculation system similar to that used on four-stroke diesels is incorporated, however, problems caused by excessive soot formation are likely to occur. Lubricating oil losses will lead to critical hydrocarbon and particulate emissions. This is one area where head-scavenged engines offer advantages. To

4.2 Diesel engines

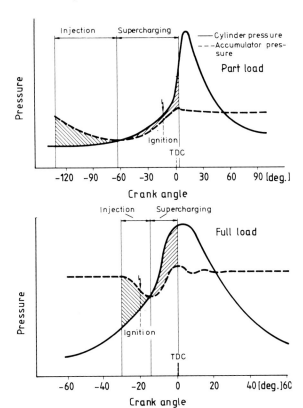

Fig. 4.13 Injection process at full throttle and part throttle in air-assisted mixture injection systems [Source: AVL]

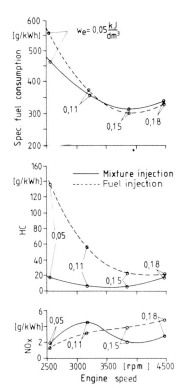

Fig. 4.14 Pollutant emissions and fuel consumption of mixture-injection and conventional injection systems [Source: AVL]

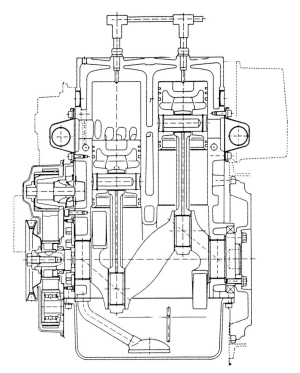

Fig. 4.15 Cross section of a loop-scavenged two-stroke diesel engine [Source: AVL]

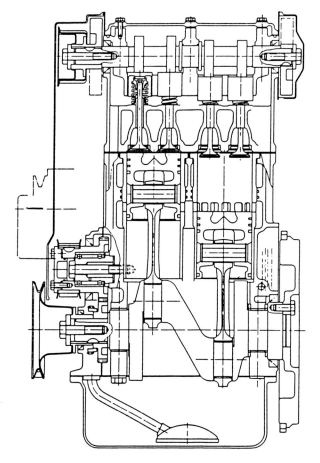

Fig. 4.16 Two-stroke diesel engine with uniflow scavenging [Source: AVL]

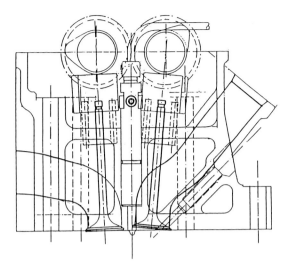

Fig. 4.17 Cylinder head of a four-valve engine with head scavenging [Source: AVL]

summarize the above statements, emission criteria currently do not yet suggest two-stroke diesel engines as a viable alternative to four-stroke diesel engines [4.2]. Two-stroke diesel engines, however, have a better chance than two-stroke spark-ignition engines of becoming available in a competitive design in the future since the combustion process is well-known. It will be adopted from four-stroke diesel engines operated with quality management. If low NO_x emissions and high exhaust gas temperatures can be achieved, emission control will also reach comparable levels.

5 Exhaust aftertreatment methods

Aftertreatment of engine exhaust emissions as a means of emission control requires different systems for spark-ignition and diesel engines. While three-way catalytic converters are used with conventional spark-ignition engines, diesel engines are fitted with oxidation catalysts and will, in the future, probably also feature particulate traps. Denox or lean-burn catalytic converters that are currently under development would be suitable both for spark-ignition engines (lean-burn engines) and for diesel engines. Lean-burn catalysts offer acceptable conversion efficiencies, especially of NO_x, in the $\lambda > 1$ range.

5.1 Gasoline engines

At present, three-way catalytic converters are the most efficient way of reducing specific exhaust components by a significant degree. Several additional methods are also available to reduce CO, HC and NO_x emissions [5.1] [5.2] [5.4]. This may be accomplished in two different ways, i.e. with thermal exhaust reactors and with catalytic converters as described in the following sections.

5.1.1 The effect of post-combustion reactions in exhaust systems

Post-combustion reactions are enhanced by implementing insulating measures to reduce heat losses and by injecting secondary air. Post-combustion reactions occur in the engine exhaust passages and involve CO and HC, in particular. To allow these reactions to occur, a sufficient temperature level and oxygen content is required. The additional oxygen required can be supplied by admixing air into the exhaust port. The point of introduction should preferably be located close to the exhaust valve since this is where exhaust temperatures are highest when the exhaust valve opens. The lower the exhaust temperatures, the more will the post-combustion reaction conditions be impaired due to reduced reaction speeds.

The required quantity of air can be supplied by an engine-driven or an electrical pump. If electrical pumps are used, speed control is omitted. Another possibility is to use self-induction by utilizing pressure pulses in the exhaust system. With this process, non-return valves cause air to be introduced via pulsations in the exhaust system.

Nitrogen oxide emissions cannot be influenced by this type of post-combustion reactions. The above measures have been adopted to an increasing extent in recent times, mainly in order to meet US emission standards on larger-displacement engines. *Figure 5.1* shows some ways of feeding secondary air into the cylinders.

Figure 5.2 shows a characteristic map of possible secondary air volumes used for medium-sized engines.

If secondary air injection is to be used for engines with $\lambda = 1$ control, this only makes sense prior to the start of λ control (e.g. in the warm-up phase) or in ranges that do not offer full $\lambda = 1$ control (e.g. at full throttle).

5.1.2 Thermic reactors

The principles of operation of thermal reactors are similar to those of the system described in Section 5.1.1. An extended residence time also is desirable as this will yield significant

5.1 Gasoline engines

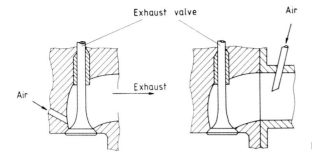

Fig. 5.1 Designs of secondary air injection [5.3]

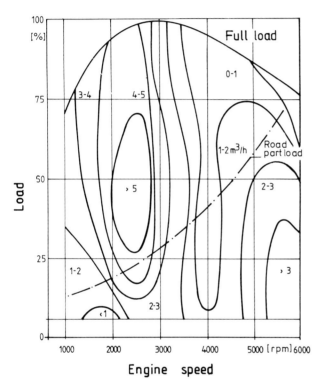

Fig. 5.2 Secondary air characteristic map [5.3]

reductions of CO and HC emissions. For this purpose, a reaction chamber with thermal insulation is used that should, if possible, be fitted directly behind the engine exhaust. Again, secondary air injection is required here if the engine is operated at rich settings.

Temperatures in excess of 700 °C inside the thermal reactor produce high conversion levels. *Figure 5.3* shows the effect on the CO and HC emission components. In this range, nitrogen oxide emissions cannot be influenced any more than in the case of secondary air injection.

Rich settings lend themselves particularly well to achieving considerable reductions of HC and CO emissions. Yet thermal reactors have failed to gain widespread acceptance. This is explained by the following reasons:

- Additional expenditure for high-temperature reactor materials and insulation, increased mass, and requirement for secondary air pump,
- Increased space requirements,
- Reactors are very effective at "rich" settings. In order to keep fuel consumption down, however, this range should not be used (at least not in the part-throttle range).

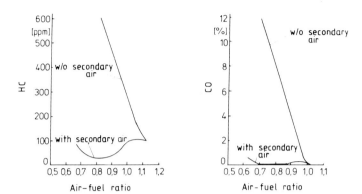

Fig. 5.3 Reduction of CO and HC in thermal reactor [5.1]

- Danger of burn-through in case of ignition failure (fuel mixture enters reactor directly),
- Insufficient conversion efficiency at low temperatures (part load),
- Virtually no effect on NO_x emissions.

5.1.3 Catalytic converter systems

Figure 5.4 shows an overview of available systems. Apart from being subject to temperature constraints, the conversion effect of catalytic converter systems depends mainly on the catalytic effect of the materials used.

Four basic designs have been developed: Oxidation, dual-bed, Denox (lean-burn) and three-way catalytic converters.

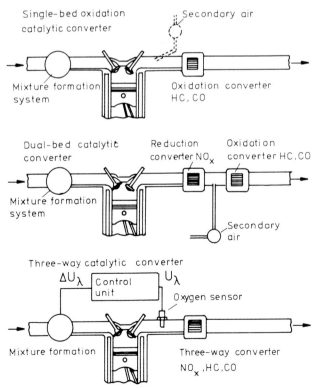

Fig. 5.4 Catalytic converter systems [5.1]

Oxidation catalytic converters: This system ideally converts the oxidising components into H_2O and CO_2. The catalyst operates at excess air settings. The air required for the oxidation process is supplied by lean mixture settings or by secondary air injection. No complex control systems are required. Oxidation catalysts were used for the first time in 1975 on U.S. vehicles. They are no longer used on passenger vehicles today since three-way catalytic converters are far more efficient and will additionally convert NO_x emissions. Oxidation catalysts are used to an increasing degree on diesel engines, however, since they are capable of oxidising soluble particulate matter in addition to the above components [5.46].

Dual-bed catalytic converters: This system includes two catalyst systems mounted in line in the exhaust system. A reduction catalyst is fitted to minimize NO_x emissions, and an oxidation catalyst is used to reduce HC and CO emissions. The engine must be operated at air deficiency ($\lambda < 1$). This system has therefore certain drawbacks in terms of fuel consumption and CO_2 emissions.

Complex mixture formation control systems are not required. Since the engine is operated in the rich λ range, increased ammonia emissions may result. The nitrogen oxide conversion efficiency is far lower than with three-way catalytic converters.

Three-way catalytic converters: A characteristic feature of this system is that is reduces NO_x, and HC and CO by the same high degree throughout. To achieve optimum emission control results, however, a complex control system (lambda control) is required.

Uncontrolled systems with three-way catalytic converters of the type that was also marketed in Germany only reach conversion efficiencies of approx. 40 to 50%, whereas computer-controlled systems in new condition will reach conversion efficiencies of more than 95%. Three-way closed-loop catalytic converters currently are the most efficient emission control systems available for internal-combustion engines.

Denox or lean-burn catalytic converters: Lean-burn or Denox catalytic converters allow not only CO and HC, but also NO_x to be converted in the excess-air range. They are currently in the development stage. NO_x conversion efficiencies in excess of 50% have already been demonstrated.

5.1.3.1 Working principles of catalytic converters

A chemical reaction can only be started if a certain energy threshold (activation energy) is overcome. The operating principle of catalytic converters is based on lowering this energy threshold far enough to allow oxidation and reduction processes to be started.

Reaction speeds are increased by this process. The temperature required to start the reactions is reduced significantly. Prior to the reaction proper, the oxidising components and oxygen are stored in the catalytically active zones. This ensures that far less activation energy is required to start the reactions. The catalytic converter primarily affects emissions of the following exhaust gas components:

- HC (unburned hydrocarbons)
- CO (carbon monoxide)
- NO_x (nitrogen oxide)

This applies to three-way, dual-bed and to lean-burn catalytic converters.

Several other reactions also occur at the same time, e.g. reactions involving sulphur, NH, NHC and OHC.

The following reactions determine HC and CO conversion as a gross conversion equation [5.5]:

$$H_nC_m + (n/4 + m)O_2 \Leftrightarrow mCO_2 + (n/2)H_2O,$$
$$H_nC_m + 2mH_2O \Leftrightarrow mCO_2 + (n/2 + 2m)H_2,$$
$$CO + 1/2\,O_2 \Leftrightarrow CO_2,$$
$$CO + H_2O \Leftrightarrow CO_2 + H_2.$$

NO_x conversion is determined by [5.5]:

$$CO + NO \Leftrightarrow 1/2\,N_2 + CO_2,$$
$$C_mH_n + (2m + n/2)NO \Leftrightarrow (m + n/4)N_2 + n/2\,H_2O + mCO_2,$$
$$H_2 + NO \Leftrightarrow 1/2\,N_2 + H_2O.$$

The first four reactions are oxidation processes and determine conversion of unburned hydrocarbons and carbon monoxide into carbon dioxide and water.

It is evident that a reduction of the CO and HC pollutants leads to an increased CO_2 content.

The reactions involving NO are called reduction processes. They determine the decomposition of NO and NO_x, respectively.

5.1.3.2 Catalytic converter design

Three-way catalytic converters are used virtually exclusively on spark-ignition engines today since they allow the three HC, CO and NO_x components to be converted. The above reactions occur simultaneously in this type of converter. The main subassemblies of the catalytic converter are:

A) Substrate material,
B) Intermediate coat (washcoat),
C) Catalytically active coat,
D) Support and housing.

A) Substrate: Ceramic and metal-based substrates are used.

a) Ceramic substrates: The substrate material (monolith) is a honeycomb ceramic body that is usually manufactured by extrusion molding processes. The honeycomb structure may be of different shapes (e.g. round, oval, rectangular) and is aligned in the direction of flow.

The number of cells is generally indicated per square inch. Usual cell numbers are 200 to 500 cells per square inch. To cope with the high exhaust temperatures attained during engine operation, this ceramic material must be highly temperature-resistant. If consists mostly of cordierite (Mg, O, Al, Si) and must have the following properties:

- High mechanical and thermal rigidity,
- Immunity to thermal shocks as it is subjected to thermal threshold loads.

The thickness of the individual cells must be kept within suitable limits to provide a sufficient flow cross-sectional area. This is essential to avoid excessive exhaust backpressure with its negative effect on engine operation (power, fuel economy).

Catalytic converters with up to 400 cells per square inch have proven to be a good compromise for practical use. The cell wall thickness inside the monolith is 0.3 mm. The

5.1 Gasoline engines

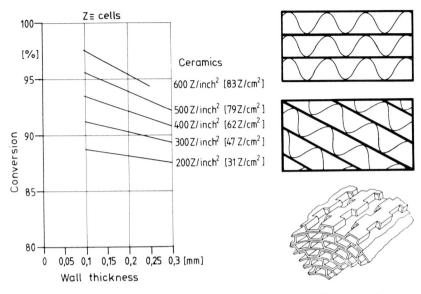

Fig. 5.5 Catalytic converter activity as a function of cell number and wall thickness [5.7]

Fig. 5.6 Structures of a metal substrate [5.50] [5.52]

temperature of the ceramic substrate must not increase significantly beyond 900 °C since this would result in impaired conversion efficiencies in long-term operation.

This is due to the fact that the effective surface is reduced by sintering processes. *Figure 5.5* shows how a change of the inner substrate geometry is reflected by a reduction of the efficiency of the catalytic converter [5.7].

b) Metal substrates: Metallic materials were introduced as support materials some time ago [5.8]. This development was sparked by advantages such as

- Increased conversion efficiencies,
- Longer life,
- Reduced installed space thanks to reduced wall thickness, or lower exhaust back pressure if installed space is kept unchanged [5.48],
- Special mounting in the housing may be omitted,
- Faster lighting off thanks to reduced mass,
- Higher permissible exhaust temperature.

A disadvantage is faster cooling at lower loads since catalyst mass is lower and heat dissipation properties are improved. For this reason, metal-substrate catalytic converters should be fitted as close to the engine as possible.

Metal-support catalytic converters feature a "monolith" consisting of a metal element (heat-resistant ferritic steel). A sheet material with a thickness of 0.04 to 0.06 mm thickness is formed e.g. as a corrugated strip, wound and connected in a high-temperature soldering process. Depending on the shape of the strip band, various honeycomb structures are possible. The honeycomb structure has a decisive influence on the mechanical properties. *Figure 5.6* shows some possible structures of a metal support [5.8] [5.9].

The main reaction zone inside a catalytic converter obviously generates the highest temperatures. The maximum temperature is reached in an axial direction of flow approx. 30 to 40 mm after the point of entry into the monolith.

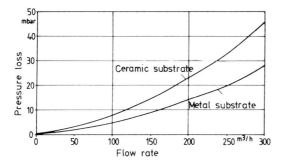

Fig. 5.7 Comparison of pressure losses of ceramic and metal substrates of equal size [5.50] [5.51]

The reduced wall thickness of the metal substrate increases the free flow area of the catalyst, thus producing lower pressure losses. *Figure 5.7* compares the pressure losses of equally sized catalytic converters with metal and ceramic substrates, respectively.

The ferritic chrome steels used allow temperatures in excess of 1,000 °C to be controlled (This, however, does not apply to the coating). The temperature resistance of the matrix can be increased further by alloying small quantities of yttrium (0.3%). No special demands are placed on the mounting of metal-substrate catalytic converters since the heat-expansion properties of the mounting are similar to those of the housing and since their sensitivity to temperature shocks is far lower.

B) Washcoat: The washcoat is applied to the substrate material. It usually consists of Al_2O_3 and is used to improve the catalytic activity of the precious-metal coating. Unlike the substrate material, it has a very large specific surface (15 to 25 sq. meters per cu. centimeter of catalyst surface). Its main purpose is to increase the oxygen storage capacity and to enhance key reactions.

C) Catalytically active coat: This coat is applied to the washcoat and consists of precious metals (platinum, palladium, rhodium). The three-way catalytic converter uses platinum and rhodium at a ratio of approx. 5:1, with the overall precious metal content per catalyst being approx. 40 to 50 grams per cu.ft. (3 to 7 grams of platinum, 0.5 to 1.5 grams of rhodium) [5.6].

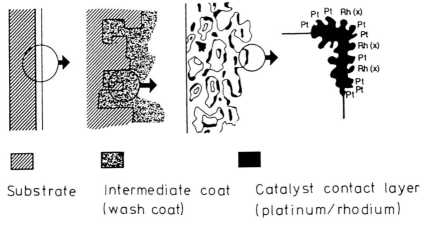

Fig. 5.8 Design of a catalytic converter (schematic) [5.44]

5.1 Gasoline engines

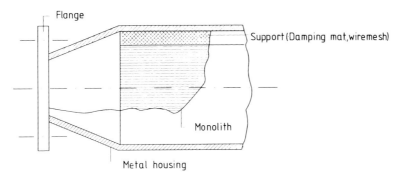

Fig. 5.9 Design of a catalytic converter system with ceramic monolith

While platinum mainly induces oxidation reactions to occur, nitrogen oxides are reduced mainly thanks to the presence of rhodium. The schematic structure of substrate, washcoat and coating is shown in *Fig. 5.8*. Initial promising results have also been obtained with coatings that are not based on precious metals.

D) Support and housing: The coated monolith is mounted in a metal housing that is part of the exhaust system. Since the ceramic material is extremely brittle and is subject to different heat expansion rates, an elastic intermediate layer to the metal housing is required. A damping mat or wiremesh is used for this purpose, with damping mats gaining increasing importance since in addition to providing better sealing, they also serve as heat insulators. The damping mat consists of ceramic fibers (aluminum silicate) and expands under the influence of heat. As the temperature rises, the surface pressing increases and thus improves the sealing effect to prevent undesirable bypass losses of exhaust gas [5.24]. *Figure 5.9* shows a schematic view of this type of system design.

5.1.3.3 Conversion efficiency

Figure 5.10 shows the effect of a three-way closed-loop catalytic converter on exhaust pollutants.

The conversion efficiency is an indicator to which extent a pollutant component in the exhaust is reduced by the catalytic converter. 100% conversion efficiency means that

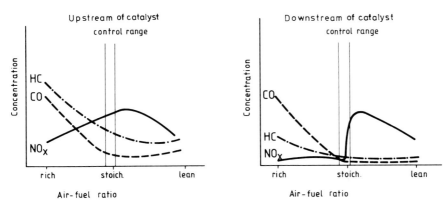

Fig. 5.10 Effects on pollutant components of a three-way closed-loop catalytic converter

a pollutant is reduced completely; 0%, on the other hand, means that the pollutant is not reduced. For any given catalytic converter, this efficiency essentially depends on two factors:

- Exhaust temperature,
- Air/fuel ratio.

The effect of both factors is dealt with in the below section.

A) Light-off performance: The light-off performance of a catalytic converter indicates the conversion efficiency as a function of the exhaust temperature. The light-off temperature may be defined as the temperature at which a 50% conversion efficiency is obtained. One of the aims of catalyst design is to keep light-off temperatures as low as possible. *Figure 5.11* shows typical conversion efficiencies as a function of exhaust gas temperature.

Depending on pollutant components and catalytic converter design, the 50% threshold is reached at a temperature between 260 and 280 °C. As is shown, CO and NO_x conversion efficiencies of approx. 90% will be reached at temperatures of 300 °C.

Especially in the first seconds of the exhaust emission tests, i.e. when the catalytic converter has not yet reached its operating temperature, the conversion efficiencies are extremely low. With regard to the applicable US emission standards (refer to Chapter 8 for a definition of TLEV, LEV and ULEV), special efforts will be required to improve conversion efficiencies, especially of unburned hydrocarbons.

One measure to solve this problem is to install a catalytic converter that is heated during the starting and warm-up phases [5.10] [5.30] [5.38] [5.39]. The following approaches may be considered:

- Electrical energy,
- Burner mounted ahead of catalytic converter,
- Afterburning chamber mounted ahead of catalytic converter.

If electrical heating is used, the battery energy will be supplied directly to the catalytic converter. The combustion gas required for heating, however, must be generated by a suitable system.

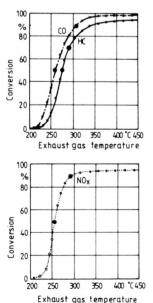

Fig. 5.11 Conversion efficiency as a function of exhaust temperature [5.5]

5.1 Gasoline engines

	NMHC	CO	NOx
Ceramic catalytic converter	0.21	2.83	0.45
Metal-substrate catalytic converter (unheated/without secondary air)	0.12	1.13	0.22
Metal-substrate catalytic converter (heated/with secondary air)	0.03	0.35	0.22

Fig. 5.12 Exhaust emissions obtained with a heated catalytic converter [5.10]

In systems based on electrical heating, the metal substrate itself acts as an electrical resistor, with only a certain segment of the active catalyst surface being heated externally. The benefits of such a system are evident in the starting and warm-up phases of the engine. The first 1 to 2 minutes after a cold start have a decisive influence on meeting exhaust test standards since the engine generates approx. 70% of the total emissions of the test cycle in this phase. Pollutant emissions may be reduced appreciably by heating the catalyst and incorporating an additional secondary air injection system. Investigations carried out on heated systems in US emission tests produced the results shown in *Fig. 5.12*. In this type of test, the catalytic converter was heated for 15 seconds before and 30 seconds after cold starting as well as 5 seconds before and 10 seconds after the subsequent hot start. Up to the present, however, German road traffic legislation does not authorize the use of preheating systems. Even a ceramic-based catalyst may be heated e.g. by fitting an electrical heater unit ahead of the catalyst.

Depending on the system layout, the installed power is up to 5 kW.

Another design focuses on heating the catalyst by injecting and combusting fuel [3.35] or by starting additional reactions upstream of the catalytic converter [5.31]. Both the catalytic converter and the oxygen sensor will reach their operating temperature after a few seconds (approx. 15 s), thus allowing pollutant emissions to be reduced at the same time. The high HC peaks, in particular, that occur during the first seconds of the exhaust test can thus be reduced by a significant margin. Burner output is approx. 15 kW. Burner operation, however, is impaired by high exhaust backpressure in the upper load and rpm range.

The third possibility is a relatively economical one. A combustion chamber incorporating the ignition unit is fitted ahead of the catalytic converter. To achieve sufficient exothermic conditions, the mixture inducted by the engine has to be extremely rich ($\lambda = 0.55$). The air required for this process is supplied by a blower. The extremely rich mixture may affect engine operation, however (e.g. ignition reliability, repeated starting).

Within the scope of the efforts to reduce emissions below the above emission standards, systems have been developed that facilitate adsorption of hydrocarbons during the first seconds of the exhaust test [5.40] [5.41]. Ideally, desorption does not occur until the catalytic converter has reached a temperature that will ensure satisfactory conversion efficiencies. Research results have shown that HC emissions can be reduced by up to 50% in this way [5.42].

B) Conversion behavior: Conversion behavior refers to monitoring conversion of the CO, HC, NO_x components as a function of the air/fuel ratio. This leaves a relatively narrow range in the vicinity of $\lambda = 1$ that produces sufficiently high conversion efficiencies (*Fig. 5.13*).

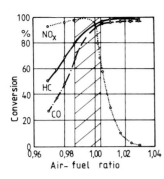
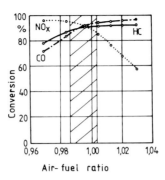

Fig. 5.13 Static conversion behavior as a function of the air/fuel ratio [5.3]

Fig. 5.14 Dynamic conversion behavior vs air/fuel ratio [5.5]

Outside the range near $\lambda = 1$, conversion efficiencies are very low, especially those of CO and HC at $\lambda < 1$ or of NO_x at $\lambda < 1$. The diagram in *Fig. 5.13* shows the static conversion behavior. This behavior is determined by adjusting a specific operational point that allows cyclic fluctuations to be kept as small as possible.

The dynamic conversion behavior is of greater importance to vehicle operation. For his purpose, conversion efficiency is determined for a pre-set average air/fuel ratio with predefined fluctuations. The effect of exhaust emission control, however, is not quite as good as in the above case. On the other hand, the permissible λ window that does not cause the conversion efficiency to be reduced dramatically is somewhat larger (*Fig. 5.14*).

The interrelationships described above refer to a "new" catalytic converter. Conversion efficiency, however, gradually decreases during the operating life of the catalytic converter. The main reasons for this are thermal ageing and "contamination".

a) Thermal ageing: This type of ageing is a process that leads to a loss of active surface of both the washcoat and the precious-metal coat due to sintering effects. In this process, the coating agglomerates and forms particulates of continuously increasing size.

Reduction of the active surface starts at the inlet of the catalytic converter. At the same time, the size of the precious-metal crystals increases, with the active precious metal surface area being reduced further (due to inclusion of precious-metal particles). *Figure 5.15* shows the loss of active surface area when the catalytic converter overheats. As shown in the diagram, the loss of specific surface area is particularly severe in the first third of the

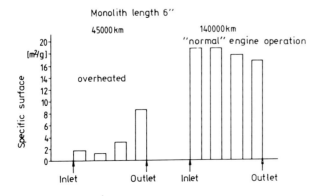

Fig. 5.15 Specific surface area in the case of overheated catalytic converter and of normal engine operation [5.6]

catalytic converter. This is the section that includes the main reaction zone of the catalytic converter.

To avoid thermal ageing of the catalytic converter, it is important to make sure that the operating temperature of the catalytic converter does not exceed specific limits during continuous operation. This limit is approx. 900 °C for ceramic catalytic converters. The longer the converter has to operate at extremely high temperatures, the more will it be affected by thermal ageing.

b) "Contamination": "Contamination" means that (physical) deposits or chemical reactions render the catalytically active surface partially inactive. This may be due to one of two reasons:

- Chemical "contamination"
  This includes chemical reactions with foreign substances such as fuel or oil additives.
- Mechanical "contamination"
  The active coat is covered (pores are closed) by additives such as lead and sulphur from fuel or by metal composites such as zinc, magnesium and calcium introduced by the engine oil.

The effect of ageing on the conversion efficiency or on the light-off performance, respectively, is shown in *Fig. 5.16* for one possible case.

Figure 5.17 shows the effect of different temperatures in the catalytic converter [5.6]. In the range near 800 °C, sintering effects involving engine oil additives and fuel can occur. A further temperature increase results in an excessive loss of surface area of the washcoat. If the threshold of 1,000 °C is exceeded, sintering effects of precious metals occur and the washcoat tends to detach itself. Temperatures that remain high for longer periods of time are particularly critical, whereas brief temperature peaks are less problematic [5.36].

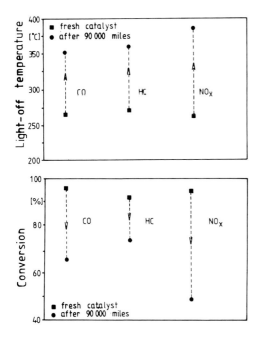

Fig. 5.16 Effect of ageing on conversion efficiency and light-off temperature at equal boundary conditions

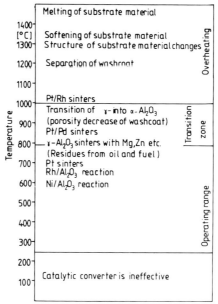

Fig. 5.17 Effect of different temperatures inside the catalytic converter [5.6]

Ageing of the catalytic converter also modifies the λ window that provides optimum conversion efficiency.

5.1.3.4 Location in the vehicle

The location of the catalytic converter under the vehicle should be selected in such a way that the following criteria are met:

- Favorable position in the flow, i.e. pressure losses of exhaust gas fed to the catalytic converter should be as low as possible,
- Equal radial supply of exhaust gas into the catalytic converter,
- Excessive cooling of exhaust gas between engine and catalytic converter must be prevented in order to ensure early light-off.

The catalytic converter should be placed as close to the engine as possible. This is particularly important in the case of metal-support catalysts as they have a lower heat storage capacity and can tolerate higher temperatures. If the catalytic converter is installed farther away from the engine, it may have to be insulated against heat losses in order to avoid cooling off in the lower load range. It should be made sure, however, that the permissible catalytic converter temperatures are not exceeded. The catalytic converter can be mounted closer to the engine in countries with vehicle speed limitations since the maximum exhaust temperatures are lower under these driving conditions.

Engine management has to be used to prevent unburned fuel from entering the exhaust system as the fuel might otherwise ignite in hot sections of the catalytic converter, generating excessive temperatures in the process. Overheating may occur, for example, if engine rpm is limited by switching off the ignition instead of by interrupting fuel supply. Further problems may be caused by ignition failure or misfire in one or several cylinders.

In certain cases, start-up catalytic converters are used that require relatively little installation space and are fitted as close to the exhaust manifold as possible or even inside the manifold. Thanks to the high temperatures at the engine exhaust, the startup catalytic converter will thus light off faster. The resulting exothermic reactions supply exhaust gas at higher temperatures to the main catalytic converter. The startup catalytic converter must be designed in such a way, however, that only a small portion of the exothermic potential contained in the exhaust gas is reduced since otherwise the catalytic converter may overheat. It therefore makes sense to use a metal substrate that can sustain higher thermal loads and requires less installed space.

5.1.3.5 Disadvantages

In addition to the problems described above, i.e. reduced efficiency due to contamination and thermal ageing, the following disadvantages have to be taken into account:

- Complex mixture control system is required to allow the air-fuel mixture to be controlled within extremely narrow limits,
- The three-way closed-loop catalytic converter has to be supplied with a stoichiometric mixture. This precludes the possibility of operating the engine at lean settings in the part-load range in order to improve fuel economy (i.e. fuel consumption is approx. 8% higher),
- Higher exhaust backpressure has detrimental effect on knock tendency, fuel consumption and power,
- Higher cost due to additional systems and use of precious metals.

5.1.3.6 Critical operating range

Excessive temperatures impair the efficiency of the catalytic converter. Depending on the operating range that the engine is operated in, exhaust temperatures may rise well beyond 900 °C. This, however, has to be prevented by suitable measures. The overrunning or deceleration mode is the most critical operating range, i.e. if no deceleration fuel cut-off is provided. *Figure 5.18* shows this range of the map, and *Fig. 5.19* shows catalytic converter temperatures that may be reached under these conditions.

As pointed out before, the permissible boundary temperature of a ceramic-substrate catalytic converter is approx. 900 °C. Damage to the catalytic converter cannot be ruled out safely at higher temperatures. The direct changeover from full throttle into deceleration is particularly critical (as it generates high temperatures).

5.1.3.7 Recycling

Worldwide use of catalytic converters is only possible in the long run if the required raw materials are available in sufficient quantities and at low cost. Attention focuses primarily

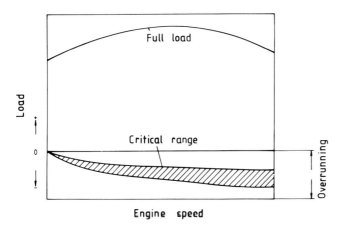

Fig. 5.18 Critical range of the catalytic converter characteristic map [Source: VW]

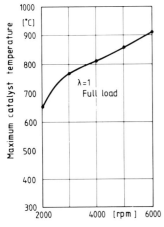

Fig. 5.19 Catalytic converter temperatures as a function of the engine speed

on precious metals, i.e. platinum and rhodium. On an average basis, each catalytic converter contains approx. 3 to 7 grams of platinum and 0.5 to 1.5 grams of rhodium. The platinum quantity required for manufacturing automotive catalytic converters currently already amounts to as much as approx. one third of the entire quantity produced. With a usage of approx. two thirds of the quantity produced, the situation for rhodium may already be considered critical [5.23].

Since the percentage of vehicles equipped with catalytic converters will continue to increase worldwide, suitable methods for recycling the precious-metal content contained in used catalytic converters should be devised. Several hundred tons of used catalytic converters are expected for 1993, with this quanitity increasing to 1,500 tons by the end of this century. This would yield a recovery of approx. 2,250 kg of platinum and 450 kg of rhodium.

For recycling purposes, ceramic and metal support catalytic converters have to be treated separately. Ceramic monoliths have to be separated from their housing and then have to be finely ground. Two alternatives are available for further treatment:

- Pyrometallurgic treatment,
- Wet-chemical treatment.

In the pyrometallurgic process, the ground ceramic material is molten. The substrate material forms slag in this process, allowing the metal elements to be separated. In a subsequent process, metals are separated from precious metals in an oxidation process. Additional chemical processes yield precious metals of very high purity.

In the wet-chemical process, oxidising acids or an alkaline disintegration process are used to separate the materials. Subsequent processing steps to yield higher levels of purity are similar to pyrometallurgic processing.

Recycling of metal-substrate catalytic converters is more complicated. The materials containing the precious metal content have to be separated from the substrate metal in an initial processing step. One promising possibility would be to machine the metal substrate. This process allows the washcoat to be detached, and the precious metals can be transferred into a gaseous phase and separated by using suitable processes.

5.1.3.8 Lambda sensors

The lambda (λ) sensor is the most important component of the entire three-way catalytic converter system. It is used to measure the air/fuel ratio (λ) and to control it exactly to a setting of $\lambda = 1$. If a three-way catalytic converter is to be used, the air/fuel ratio must remain within a narrow λ window ($\lambda = 0.995$ to 1.005) so as to ensure high conversion efficiencies (*Fig. 5.20*). The λ sensor is located in the exhaust system in a position that provides an exhaust gas composition representative of all cylinders. V-engines sometimes use two sensors.

The operation of the sensor is based on the principle of a galvanic oxygen concentration cell with a solid electrolyte. The ceramic material of the sensor becomes conductive from approx. 300 °C.

Figure 5.21 shows the basic principle of this sensor type.

A mixture of zirconium and yttrium oxide is used as a gas-permeable solid electrolyte. This mixture is a virtually pure oxygen conductor. If both sides of the solid electrolyte are bonded with porous electrodes and if one side is exposed to a higher oxygen concentration than the other side, an electrical voltage is obtained. This voltage may thus be used as an indicator of the oxygen concentration in the exhaust. The oxygen contained in the engine exhaust emissions is directly dependent on the λ setting. As a result, the oxygen content can

5.1 Gasoline engines

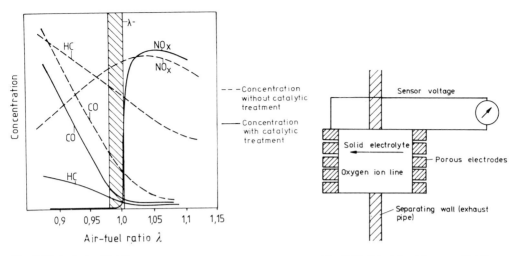

Fig. 5.20 λ window [5.14]

Fig. 5.21 Principle of λ sensor [5.14]

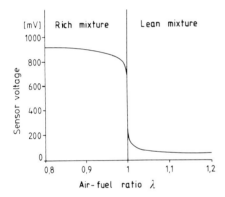

Fig. 5.22 Voltage curve of a λ sensor at 600 °C exhaust temperature [5.14]

be used to reflect the air/fuel ratio. *Figure 5.22* shows the sensor voltage output as a function of the oxygen content in the exhaust. It is evident from this diagram that the characteristics of the sensor voltage allow particularly accurate correlations to be made within a very narrow λ range. This response characteristic is called a step function and is used for λ control.

In addition to the oxygen content in the exhaust, the temperature of the ceramic body is another important parameter since it has an effect on oxygen ion conductivity. The output voltage is influenced to a great extent by temperature. The optimum operating temperature is approx. 600 °C, minimum temperature is 300 °C.

When the engine is started, the minimum operating temperature has not yet been reached and the closed-loop control system is therefore switched off. The engine is now controlled in an open-loop mode. After the desired temperature has been reached, the control circuit is switched on. The λ sensor must not be operated at temperatures above 900 °C for longer periods of time.

In order to overcome problems of light-off time and voltage change due to temperature effects, e.g. after cold starting, a heated λ sensor is used. A built-in electrical heater element raises the ceramics temperature at lower exhaust temperatures so that the optimum operating temperature is always present at the sensor. This offers the following advantages:

- Faster heating (λ control can be activated approx. 20 sec after starting the engine)
- Lower and stable emissions.

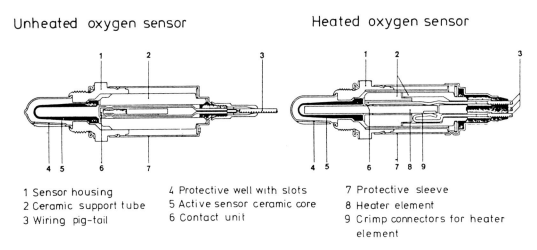

Fig. 5.23 Cross-section of a heated and an unheated λ sensor [5.14]

Figure 5.23 shows the design of a heated and an unheated lambda sensor.

To preserve the full efficiency of the catalytically active coat of the platinum electrode of the λ sensor, the engine must be operated with unleaded fuel. Deposits of lead contained in the fuel would cause the catalytically active coating to be damaged ("contaminated").

Recent studies have shown that the precision of λ control may be increased if a second sensor is incorporated downstream of the catalytic converter [5.26]. The signal supplied by this sensor is used as a guideline for λ control, thus allowing detrimental effects on the λ sensor ahead of the catalytic converter to be compensated.

Another development is based on measuring the residual oxygen content directly as an indicator of the air/fuel ratio.

Higher temperature resistance and reduced response time of this type of sensor facilitate implementation of cylinder-selective mixture control systems and thus offer benefits in terms of exhaust quality and fuel consumption [5.27]. λ sensors suitable for use with lean-burn engines have since been introduced as well.

5.2 Diesel engines

Diesel or compression-ignition engines operate at far higher air/fuel ratios than spark-ignition engines. Since engine load is controlled on the basis of the air/fuel ratio, this parameter cannot be used to influence exhaust composition. An influence is only possible if fuel supply is limited at full throttle. This allows the smoke number and soot emissions to be limited. In the past, this used to be the only parameter that was taken into account in terms of diesel emission control. This has changed dramatically since the mid-80s, though. As in the case of spark-ignition engines, far more stringent emission standards have since been introduced.

5.2.1 Thermic reactors

The operation of thermal reactors is the same with both SI and diesel engines. Sufficiently high temperatures are required to reduce CO and HC emissions. Diesel engines, however, i.e direct-injection diesel engines in particular, operate at far lower exhaust gas tempera-

5.2 Diesel engines

tures. If the full operating range of the diesel engine is taken into account, thermal reactors have only a rather limited effect. Similar facts apply to post-reactions of soot, as both the temperature and the residence time in the reactor are insufficient. Thermal reactors therefore are even less suitable for diesel engines than they are for spark-ignition engines.

5.2.2 Catalytic reactors

Three-way catalytic converters of the type used for spark-ignition engines are not suitable for diesel engines since diesel engines cannot be operated at $\lambda = 1$. Due to the high excess oxygen ratio across wide operating ranges, virtually no significant catalytic NO_x reduction is present. This is due to the fact that the basic reduction potential of CO and HC is used up by oxidation processes.

Another drawback is that exhaust temperatures are not always sufficient to ensure safe light-off of the catalytic converter. Some alternative processes have been investigated but have not yet been introduced on a production scale:

- **Selective non-catalytic reaction**, with ammonia or urea being used as a reductant. These processes are used for flue gas denitrification in power plants and for stationary diesel engines.
- **Selective catalytic reduction (SCR):** Catalysts and reductants, e.g. NH_3 (ammonia) are used. The Carnox process is one of the processes that are currently under development. Dependent on NO_x emissions and catalyst temperature at a given moment, a reductant is added ahead of the catalyst via map control (*Fig. 5.24* [5.28]). Carbamide is used as a reductant and is added as a water-based solution using air flow. Initial applications have been made on large-size diesel engines.
- **Non-selective catalytic reduction (NSCR):** Conversion efficiencies are lower than with the SCR process. In addition, a specific content of oxidising components is required. These components may be supplied externally or by suitable engine-related processes (injection characteristics). If specially coated catalysts are used, HC, CO and particulate emissions can be reduced along with NO_x [5.32].
- **Denox or lean-burn catalyst:** Since both the NSCR and SCR processes primarily focus on reducing NO_x in the lean range, catalysts of this type often are referred to as Denox catalysts. This catalyst design (zeolite) [5.33] featuring on a copper-iron base (ZSM-5) has yielded NO_x emission reductions of approx. 20% in the diesel exhaust. If, on the other hand, the content of specific HC components is increased in the exhaust, NO_x reductions

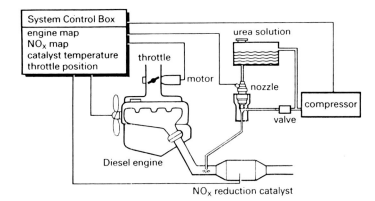

Fig. 5.24 The Carnox process

of more than 50% can be achieved in conjunction with a hot EGR system. Hydrocarbons that, at least partially, have to be introduced in a vaporized state are used as reductants in this process.

5.2.2.1 Oxidation catalytic converter

As pointed out before, oxidation catalytic converters are currently the only type of catalytic converter used for diesel passenger vehicles. They allow hydrocarbons, carbon monoxide and soluble particulates to be oxidised above a temperature of approx. 170 °C. An important prerequisite is that

- the catalyst is not "contaminated" by sulphur oxidation products,
- the catalytically active surfaces are not polluted e.g. by soot deposits.

Figure 5.25 shows the factors that affect the efficiency of oxidation catalysts. The coating is of particular importance in this context (to suppress sulphate formation and to ensure early light-off at low temperatures).

The risk of contamination and pollution is relatively small, however, since the soot particulates have been dried by oxidation and are removed by the high flow velocities. Sulphur contamination is not a huge risk, either, since this is a reversible process that causes stored sulphates to desorb again at high temperatures. The presence of SO_2, however,

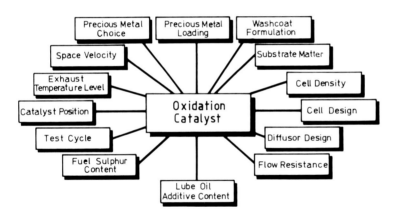

Fig. 5.25 Factors that affect the efficiency of an oxidation catalyst [5.45]

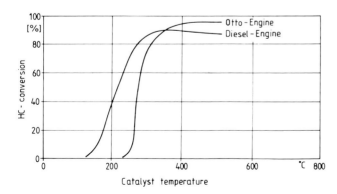

Fig. 5.26 Basic diagram of HC conversion efficiencies of SI and diesel engines as a function of temperature [5.49]

5.2 Diesel engines

reduces HC and CO conversion efficiencies. Since the sulphur content in future diesel fuels will be reduced yet further, contamination should be even less of a problem in the future.

Figure 5.26 shows a comparison of the HC conversion efficiencies of spark-ignition and diesel engines. It is evident that conversion starts at far lower temperatures on diesel engines. This is due to the high excess air ratio (good availability of oxygen) and to the composition of diesel fuels. The problems of using a catalytic converter on diesel engines may be summarized as follows:

- Pollution of catalytically active surface by soot and ash residue. This can be avoided safely by selecting the catalytic converter size and cross-section in a suitable way.
- The risk of sulphur contamination may be avoided by selecting a suitable coating.

Sulphate (SO_3) formation is restricted by using a platinum-rhodium or palladium coating. Rhodium and palladium suppress the conversion of SO_2 into SO_3.

- If an oxidation catalyst is used, the organically soluble contents of particulate emissions at part throttle are reduced by post-combustion. The sulphur contained in the diesel fuel, however, leads to increased particulate emissions due to sulphate formation in the upper load range at higher temperatures (350 to 400 °C). It would be desirable to reduce the sulphur content yet further in order to reduce particulate emissions. Suitable measures have so far only been adopted in very few countries, however. Lowering the sulphur content to 0.05%, as discussed repeatedly, is a must as this allows emissions of non-soluble particulates to be reduced. Problems of economical fuel production have to be taken into account in this context. *Figure 5.27* shows the use of non-sulphur fuel and of sulphur-containing fuel. In addition, *Fig. 5.28* shows the interrelationship between exhaust temperature and sulphate or particulate formation.

The light-off point (temperature at which a 50% conversion efficiency is achieved) of the catalytic converter should not be undercut during engine operation as "white smoke" will be emitted when the converter lights off again.

The potential for improvement that is achieved if a swirl-chamber engine is fitted with a catalytic converter is shown by the emissions measured during FTP-75 testing (*Fig. 5.29*).

As was to be expected from the above considerations, an oxidation catalyst offers significant advantages in terms of CO, HC and particulate emissions. The reduction of

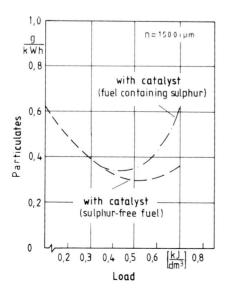

Fig. 5.27 Particulate emissions of engines with catalytic converter [5.1]

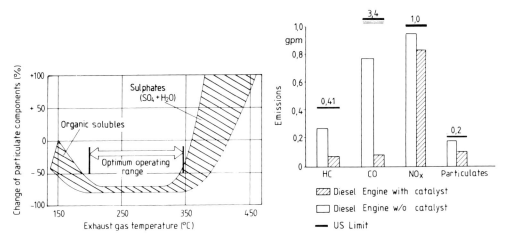

Fig. 5.28 Exhaust temperature and particulate formation [5.15]

Fig. 5.29 Emissions of a swirl-chamber engine recorded in US tests with and without catalytic converter [Source: VW]

particulate emissions, however, is largely dependent on the test cycle selected and on the sulphur content of the fuel. Depending on the time share of high exhaust temperatures, the test cycle determines the level of particulate formation (sulphate formation).

The increased use of diesel catalytic converters will also have a profound effect on engine characteristics. In the case of particulate emissions, in particular, efforts should be made to increase the share of soluble components and to reduce non-soluble particulate components at the same time since soluble components can be decomposed by the catalytic converter, thus increasing the particulate-reduction effect of the diesel catalytic converter.

5.2.3 Separation systems

The reduction of vehicle engine particulate emissions is a problem of increasing importance, especially for diesel engines. On one hand, particulate emission standards become increasingly stringent, on the other hand, current particulate traps or filters suitable for large-scale use in passenger car engines do not yet meet the requirements.

Particulate size distribution is another problem area. Many of the anti-pollution methods indicated in *Fig. 5.30* cannot be used efficiently for engine-related exhaust emission control.

The different trap (filter) systems can be differentiated according to the filter material used. Diesel particulate trap materials consist mainly of ceramics, steel, sinter metals and fibers that feature either a catalytic or a teflon coating. Exhaust gas passes the trap across a close-woven tissue or porous media (porous walls) with a small free cross-section and a large surface. The fiber spacing of wrap-around filters often exceeds the size of the most current particulate diameters by one to two powers of ten. This means that only larger particulates are retained directly by the filter whereas smaller particulates are deposited on the filter wall in an absorption process. As the filter is charged increasingly with particulates, the filter efficiency increases. This, however, results in an increased cross-sectional resistance and thus in increased exhaust back-pressure. This characteristic explains why the filter will have to be regenerated as soon as a specific filter charging condition is reached. The entire problem of particulate trap technology is evident from the conflicting interests of deposition rate/exhaust backpressure/regeneration/installation size.

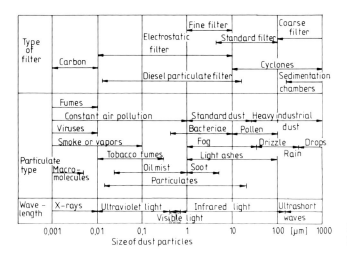

Fig. 5.30 Particulate sizes and filter types [5.18]

Both cyclone collectors, activated carbon, impingement or diffusor collectors and flue gas washers have failed to find widespread use in automotive applications. The same is true for collector systems used for industrial gas treatment. Recently tests have been carried out on collectors that agglomerate particulates using an electrostatic field and separate them from the exhaust in a downstream system. Following this process, the particulates are then recirculated for treatment by engine combustion or intermittent burnoff.

Problems of this system are relatively low trap efficiency (50%), high design complexity and increased wear caused by particulate recirculation [5.34].

5.2.3.1 Trap systems

With the perspective of large-scale production introduction, regenerative particulate traps (filters) are considered to offer the best potential. No system has yet entered production on a constant basis since no safe solutions have so far been found for the regeneration problem (burnoff of the accumulated particulate mass). Two types of trap are used:

- Surface-type traps
- Depth-type traps

Both are based on the principle of retaining the particulates in the filter bed and burning them off intermittently. Surface-type traps have a very small pore diameter. The particulates cannot enter the filter material but are deposited on the surface instead. The filter cake that accumulates on the surface of the filter material gradually clogs the trap and increases flow resistance. A large filter surface therefore is a must. Ceramic filters are the most common design used for this type of trap. The following design versions are available:

- **Ceramic monolith filter**

    The ceramic body has a multitude of ducts that are closed at alternating ends (*Fig. 5.31*).

    The gas supplied by the engine is forced to pass the porous walls, causing the particulates to deposit on the walls. As the filter cake thickens and, hence, the charging condition of the filter increases, the deposition rate increases as well.

    The advantages of such a system are: High deposition rate, low cost and reduced space requirements as well as high temperature resistance of the ceramic material. Filter efficiency depends on pore size, cell thickness and wall thickness. Ceramic monolith

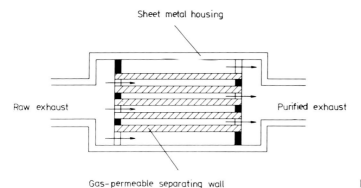

Fig. 5.31 Ceramic monolith filter

filters can be coated with catalytically active materials. The catalytic effect and the resultant exothermic processes offer benefits with regard to regeneration.

The drawbacks of such a system are a tendency of fracturing due to temperature differences (heat tension) and increased exhaust backpressure that increases even further during the service life of the system (due to the effect of deposited non-oxidising soot components).

Experiments with a trap (4.66 in. dia., 6 in. filter length, 100 cells/sq.in) show that particulate deposition rates of up to 85% can be achieved [5.6]. The required trap size is roughly equal to the swept volume of the engine. Filtering of high-boiling hydrocarbons is a problem with all particulate traps since hydrocarbons are of a gaseous nature and are transformed into particulates only when they cool off.

- **Ceramic foam filters**
Experience gained with this filter type basically is similar to the results obtained with ceramic monoliths. Due to the three-dimensional cell structure, both pore size and installation volume are larger; the deposition rate, on the other hand, is lower.

Foam filters are manufactured by saturating the filter material (Cordierite, Al_2O_3) with polyurethane.

As in the case of surface-type filters, the deposited particulates reduce the free pore cross-section and increase both the filter effect and exhaust backpressure.

- **Fiber filters**
Fiber filters have fibers with a dia. of 5 to 25 m. They are suitable for filtering diesel particulate sizes of 0.1 to 1 m. Since the fibers are extremely thin, systems of this type have only a limited service life. The larger the filter fibers for a given installation space, the lower is the filter effect and the longer is the filter life.

- **Ceramic yarn wrap-around filters**
This filter consists of perforated steel tubes that are closed at the ends and are wrapped with ceramic yarn. The exhaust gas flows from the outside to the inside, starting at the ceramic yarn layer. The particulates are deposited on this layer (*Fig. 5.32*). Deposition rate and regeneration intervals are similar to those of other systems. Improved mechanical rigidity is a major advantage of this design.

- **Steel wool filters**
Steel wool filters (*Fig. 5.33*) have filter elements with a wiremesh of approx. 0.25 mm wire thickness. They can be positioned directly at the engine exhaust port, with the filter being

5.2 Diesel engines

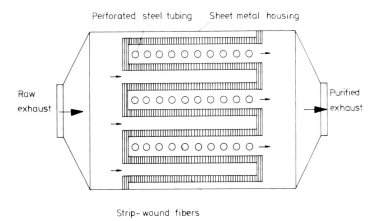

Fig. 5.32 Ceramic yarn wrap-around filter

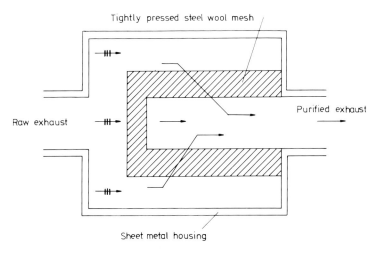

Fig. 5.33 Steel wool filter [5.13]

incorporated into the exhaust manifold. If the steel wool is given a precious-metal coating, a catalytic effect involving HC post-reactions can be achieved. Filter efficiencies are approx. 60%.

At the beginning of the charging process of the filter, only a small fraction of the particulates is retained. This rate then increases progressively. Problems with this filter type are caused by corrosion of the steel wool.

- **Sinter metal filters**
  The sinter metal filter is designed as a depth filter and consists of wire mesh layers with metal powder and sinter coatings. The filter plate produced in this process offers high mechanical rigidity and good heat conductivity. The deposition rate can be varied by modifying the thickness and porosity of the filter plate [5.29]. An additional catalytic coating can be applied on the side that comes into contact with the purified exhaust.

- **Electrostatic precipitator (electrical filter)**
  Electrical filters usually consist of an ionizing section that is used to electrically charge the particulates contained in the exhaust. Most of the particulates have a positive charge.

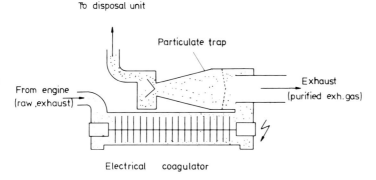

Fig. 5.34 Electrostatic precipitator [5.13] [5.14]

The particulates are then deposited across an electrical field. An advantage of this design is that (even after higher mileages) exhaust backpressure does not increase.

The particulates trapped by the filter can be reintroduced into the engine for combustion across the required exhaust gas recirculation system or can be removed in a collector tank. Recirculation systems, however, have led to increased wear in the engine. System drawbacks include increased installation space requirements, installation of a high-tension system as well as relatively low overall efficiency (deposition rate approx. 50%).

Figure 5.34 shows the basic design of such a system [5.34].

5.2.3.2 Trap regeneration

The free flow area of the trap (filter) is reduced by deposits, causing exhaust backpressure to increase. Particulate traps should therefore have a sufficient collector volume since permanent regeneration during vehicle operation is too complicated. As space constraints tend to limit the collector volume, the filters have to be regenerated at certain intervals.

Unless additional measures are applied, soot will only oxidise at temperatures above approx. 550 °C. As indicated in *Fig. 5.35*, such temperatures are reached only rarely during diesel engine operation.

This means that regeneration must be started separately after a specific interval. The following basic methods are available for this purpose:

- Burn-off of filter and filter coating by implementing engine-related measures to increase temperatures,
- Introduction of external energy to burn off the filter coating,
- Oxidizing using catalytic converters,
- Filter replacement.

The latter measure is not acceptable for motor vehicles for cost and logistic reasons.

One common feature of all other systems is that the regeneration process has to be started as soon as a predetermined charging state (exhaust backpressure) is reached. One possibility is to regenerate the filter after a specified mileage. For one thing, a minimum amount of charge is required to release sufficient energy when the soot burnoff process is started so that the deposited soot can be burnt off without having to add extra energy continuously. Excessive filter charging should also be avoided since excessive energy would otherwise be released when the regeneration process is started, causing the ceramic support

5.2 Diesel engines

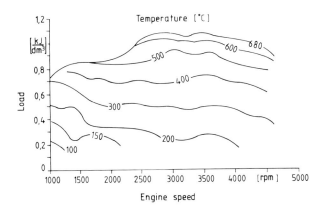

Fig. 5.35 Exhaust temperature map of a turbocharged swirl-chamber diesel engine [5.28]

material to glaze and, hence, to be destroyed. In addition, both a minimum level of oxygen content in the exhaust and a certain residence time of the soot (which may amount to several seconds depending on the air/fuel ratio and temperature) are required.

- **Regeneration by increasing exhaust temperature**
  A number of measures are available to increase the exhaust gas temperature for brief intervals. The exhaust temperature required to regenerate the filter may be obtained by combining various measures under certain operating conditions of the engine. In most cases, this will increase fuel consumption.

  – *Increasing the exhaust gas temperature by intake air throttling (IAT)*
  Throttling the intake air is a very efficient way of increasing the exhaust temperature. Intake air throttling not only increases the charge changing work but also reduces the amount of inducted fresh air considerably. As this causes the related efficiency to decrease, fuel consumption requirements increase. The parameters modified by this process increase the mixture heating value and, therefore, result in higher exhaust gas temperatures. An added effect is that throttling reduces the compression final pressure. Due to changing the thermal boundary conditions, ignition lag increases and combustion tends to be retarded. *Figure 5.36* shows a schematic diagram of the parameters that will affect intake air throttling.

  – *Exhaust gas recirculation (EGR)*
  Exhaust gas recirculation causes the temperature to increase by admixing hot exhaust gas to the intake air. The effect on exhaust temperature is only limited, however, and this approach therefore is not suitable for improving the boundary conditions for soot oxidation.

  – *Intake air preheating (IAP)*
  Preheating of the intake air raises the fresh air temperature. This reduces ignition lag. Research has shown that, especially in the part-load range, exhaust temperature increases only to a very limited extent. If adopted individually, intake air preheating is therefore not sufficiently efficient to produce any significant effect on soot oxidation. An additional drawback is that preheating the intake air increases nitrogen oxide emissions.

  – *Combination of intake air preheating and throttling (PT)*
  Higher exhaust temperatures are achieved by implementing a combination of different measures. As an example, *Fig. 5.37* highlights the effect of different measures at a filter inlet temperature of 600 °C. The effect of a given measure extends to the line drawn in *Fig. 5.37*. A positive effect is attributed in this context to combining intake air throttling

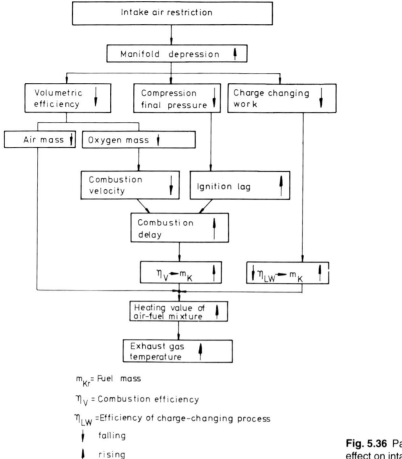

Fig. 5.36 Parameters having an effect on intake air throttling [5.20]

and preheating. Even this combination, though, leaves a map section in the lower part-load range that cannot be used for filter regeneration.

As a result, the above measures do not provide safe regeneration rates throughout.

- **Regeneration using burner systems**
  A safe way of disposing of the accumulated particulates is by using a burner. With this type of regeneration, a separate burner adds energy so that the temperature required for particulate oxidation can be reached. This, however, entails an increase in fuel consumption of approx. 3% in current systems.

  Unlimited regeneration is possible across the entire map as well as during vehicle operation.

  If combined with wrap-around filters or sinter metal filters, this regeneration system currently appears to offer the best potential for minimizing diesel particulates, especially in terms of long-term stability.

- **Regeneration with electrical energy input**
  This particulate trap system consists of two parallel monoliths that are used in an alternating manner. If a preset charging level is reached in one filter (approx. 40 g of soot), the system switches to the other monolith. The charged system is regenerated electrically

5.2 Diesel engines

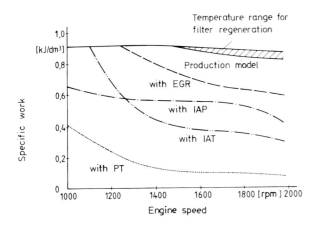

Fig. 5.37 Map areas suitable for measures to increase exhaust temperature

by adding secondary air. An electronic control system processes, among others, signals from pressure transducers, temperature sensors and air flow meters and controls a bypass flap. The filter is regenerated every 4 to 5 hours; this process requires approx. 15 minutes. Fuel consumption increases by approx. 1 to 2%, this being caused primarily by added regeneration energy and to a lesser extent by increased exhaust backpressure [5.37] [5.43].

The system is used mainly for commercial vehicles (buses). Service life expectancy of all soot burnoff filters still remains to be improved.

- **Regeneration with oxidants**
Special additives [5.47] allow the ignition temperature of soot to be reduced down to 250 °C (*Fig. 5.38*). A number of constituents based on Cu, Fe (e.g. Ferrocen), calcium, manganese and cerium are suitable for this purpose and are added to the fuel or are introduced directly ahead of the particulate trap. The required substances can be added, for instance, across a separate tank (at a rate of 6 liters per approx. 25,000 miles).

A problem of this approach, however, is that the effects of the substances that are released into the environment in this process have only been explored insufficiently, e.g. in the case of heavy metals and ferrous compounds. This type of particulate trap regeneration is currently being tested for its environmental compatibility.

- **Catalytic regeneration**
Catalytic regeneration coats the filter with a catalytically active substance, e.g. copper oxide. As soon as the particulate trap is charged with a sufficient quantity of soot, the catalytic converter is activated using an additive (such as acetyl acetone). This reduces the regeneration temperature to approx. 250 °C, allowing regeneration to occur across wide ranges of the engine operating range. These systems require a complex control system similar to that of burner-based regeneration systems. Fuel consumption is approximately equal. *Figure 5.39* shows the basic layout of such a catalytic regeneration system.

The soot combustion temperature can also be lowered by applying a vanadium oxide coating.

Most particulate filtering methods currently researched and tested are not suitable for large-scale application in passenger cars. The extreme size of such particulate traps means that they cannot be housed in passenger cars. The available systems are suitable mainly for commercial vehicle applications. The required regeneration intervals on commercial

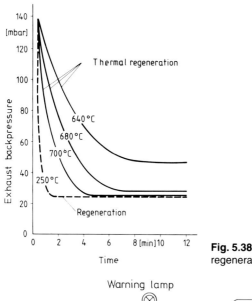

Fig. 5.38 Temperature- and time-dependent particulate trap regeneration vs exhaust backpressure [5.11]

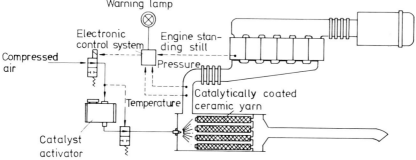

Fig. 5.39 Ceramic wrap-around filter with catalytic regeneration [5.11] [5.22]

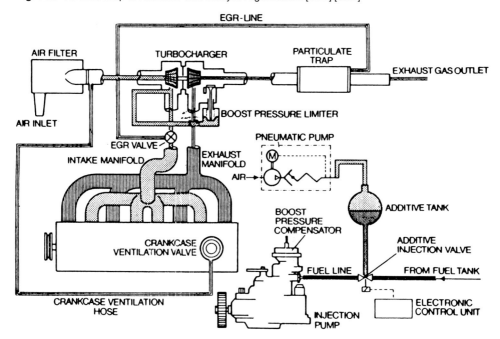

Fig. 5.40 Functional diagram of a particulate trap system [5.13] [5.28]

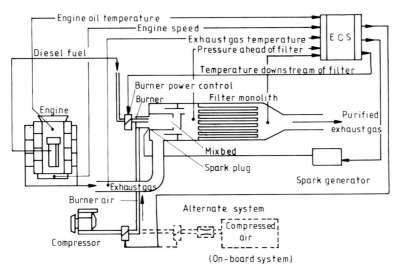

Fig. 5.41 Functional diagram of a particulate trap system [5.13]

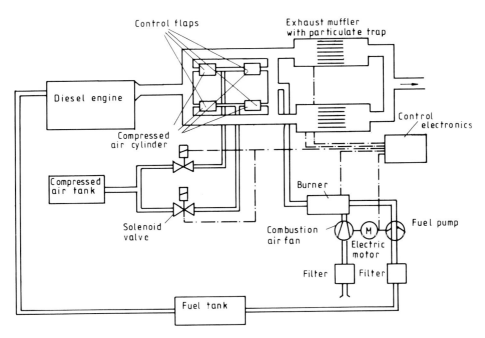

Fig. 5.42 Functional overview of a particulate trap system

vehicles currently are approx. 300 to 600 miles. For this reason, traps are therefore used primarily on public utility vehicles.

The functional diagrams of some particulate trap systems shown in *Figs. 5.40, 5.41* and *Fig. 5.42* illustrate the technological effort involved in such systems.

The only manufacturer that has so far fitted passenger car diesel engines with particulate traps for the US market was forced to withdraw the system from the market since the traps tended to clog during short-distance operation, resulting in power loss due to increased exhaust backpressure. When the vehicles were subsequently driven at full throttle, soot burnoff caused the particulate trap to be destroyed.

Since residue formation in the particulate trap causes exhaust backpressure to increase, turbocharged diesel engines may have to be operated with mechanical chargers instead of exhaust turbochargers that utilize exhaust backpressure.

If the availability of exhaust emission control systems and the required expenditure, particularly for passenger vehicle diesels, are taken into account, a diesel catalytic converter appears to be a better solution in the long run [5.46]. This statement, however, may eventually have to be reconsidered once further knowledge has been gained on the effect of engine exhaust particulates on human health.

6 The influence of fuel and lubricants on emissions and fuel consumption

6.1 Conventional fuels

The emission characteristics of internal-combustion engines are largely dictated by the type of fuel used, although not primarily in terms of the level of pollutant concentrations but rather with regard to the components contained in the exhaust. This becomes apparent e.g. when alternative fuels are used. Even with conventional fuels, however, the fuel composition is reflected by a spectrum of exhaust components typical for a particular type of fuel. This is why emission criteria will play an increasing role in the production of future conventional fuels [6.1].

6.1.1 Gasoline

Gasoline for spark-ignition engines consists of a mixture of hydrocarbons that is liquid at room temperature and covers a boiling range of approx. 30 °C to 215 °C. To improve cold starting, it is enriched with easily volatile constituents in winter.

The following criteria are of importance for exhaust emissions and fuel consumption:

- Good starting across a wide temperature range ($-40\,°C$ to $+50\,°C$),
- Good transient characteristics (load changes) and transient operation,
- Low vapor-lock tendency,
- High energy density,
- High knock resistance.

In order to meet the above criteria, specific fuel properties are required that may in certain cases have a direct impact on design and tuning of a spark-ignition engine. The main parameters are: Density, boiling curve, composition, steam pressure, heating value, octane rating, lead content, content of oxygen-based components etc. *Figure 6.1* shows some important quality criteria for spark-ignition engine fuels and their effect on engine operation.

Lead emissions have been reduced dramatically since "unleaded" fuel was introduced. The remaining lead emissions are caused by older vehicles that still require leaded gasoline. The gradual reduction of the lead content of fuels dictated by legislation standards is shown in *Fig. 6.2*.

The main substances used to improve knock resistance of leaded gasoline, i.e. lead alkyls such as tetraethyl lead (TEL) and tetramethyl lead (TML), had to be replaced by fuel components with equal anti-knock properties. For economical reasons, however, these components can only be added in limited quantities [6.3]. The octane rating of unleaded regular gasoline has remained unchanged whereas it was not possible to retain the original octane rating of premium gasoline. Following harmonization of EEC specifications, the ratings for premium gasoline now are 95 RON and 85 MON (formerly 98 RON and 87 MON). The dependence of specific fuel consumption and specific work on the RON is shown in *Fig. 6.3* [6.4]. This is explained by the higher compression potential that increases along with the octane rating.

In addition to highly knock-resistant fuel components, alcohols and esters are suitable as "lead replacements". The limit for adding alcohol is approx. 3% by volume, not least of all because of limited availability [6.2]. The use of components containing oxygen is also

	Requirements to DIN 51 600		Requirements to DIN 51 607 (unleaded)		Effect on engine operation
	Premium	Regular	Premium	Regular	
Anti-knock index RON	min 98,0	min 91,0	min 95,0	min 91,0	Knock at low or medium rpm
MON	min 88,0	min 82,7	min 85,0	min 82,5	Knock at high rpm and high load
Density at 15°C g/ml	0,730–0,780	0,715–0,765	0,740–0,790	0,720–0,770	Fuel consumption, exhaust emissions
Volatility Steam pressure bar	Summer 0,45 – 0,70 Winter 0,60 – 0,90		Summer 0,45 – 0,70 Winter 0,60 – 0,90		Cold starting (winter) Hot starting (summer)
Transient condition at 70°C Vol %	Summer 15 – 40 Winter 20 – 45		Summer 15 – 42 Winter 20 – 47		Cold starting-hot starting, Drivability with hot engine
Transient condition at 100°C Vol %	Summer 42 – 65 Winter 45 – 70		Summer 40 – 65 Winter 42 – 70		Drivability with cold engine
Final boiling point °C	max 215		max 215		Wear(cold engine operation) Residue formation
Existent gum mg/100 ml	max 5		max 5		Residue formation
Lead contents g/l	max 0,15		max 0,013		Destruction of catalytic converter and oxygen sensor

Fig. 6.1 Quality criteria for gasoline fuels (Germany) [6.2] (as of 1992; modifications of the DIN standard are planned for late 1993)

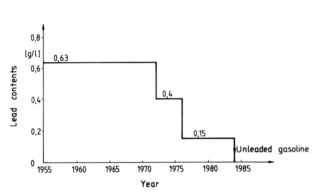

Fig. 6.2 Reduction of lead content [6.35]

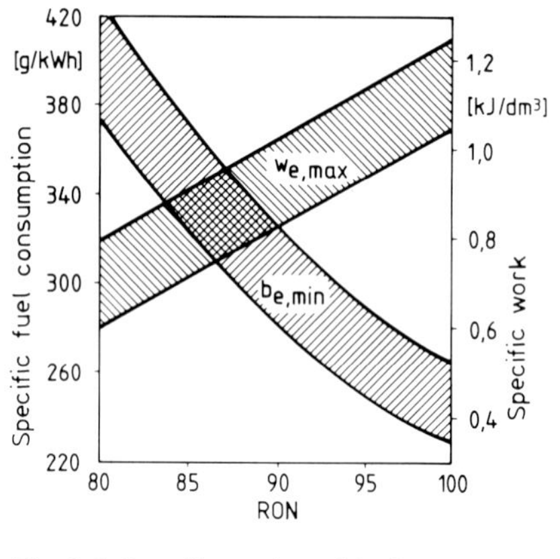

Fig. 6.3 Specific work and fuel consumption depending on RON [6.4]

critical as this increases fuel volatility and may lead to drivability problems caused by vapor locks. In addition, exceeding the admissible steam pressure limits may cause excessive evaporative losses (during refueling, across tank vent), resulting in further environmental pollution due to hydrocarbon emissions. Another critical topic is the content of benzene or aromatic substances in the fuel.

Due to environmental considerations, demands are being made to reduce the content of benzene and aromatics in the fuel yet further. *Figure 6.4* shows an example of the relationship between benzene in exhaust emissions and the benzene and toluene content in the fuel. These measurements were made on a single-cylinder engine. Fuel mixtures containing benzene or toluene and isooctane show a linear relationship with the benzene and toluene content in the exhaust emissions [6.4] [6.5].

6.1 Conventional fuels

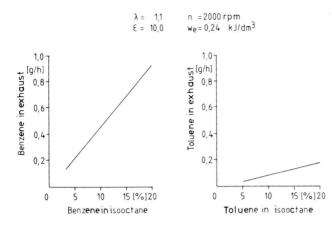

Fig. 6.4 Interrelationship between benzene in fuel and in exhaust emissions [6.5]

6.1.1.1 Reformulated gasoline

In response to the effect of fuel composition on emissions, an amendment to the "Clean Air Act" was enacted in the USA in 1990 [6.9] [6.10] that requires measures to reduce air pollution by 1995 in all heavily polluted urban areas that do not meet the national air quality standards. One measure to reach this goal is the introduction of reformulated gasoline. This term refers to fuel that has a benzene content < 1% by volume, a total of aromatic compounds < 25% by volume and an oxygen content (e.g. as contained in alcohols) < 2% by volume.

The amendment to the "Clean Air Act" additionally stipulates that, if reformulated gasoline is used, emissions of four compounds and compound classes considered to be toxic have to be reduced by 1995 by 15% over the 1990 levels and, if this is technologically feasible, have to be reduced by 25% by the year 2000. These four toxic compounds are:

- Formaldehyde,
- Acetaldehyde,
- Polycyclic organic compounds,
- Benzene.

Tests to FTP-75 have shown that the use of reformulated gasoline allows emissions to be reduced to an extent that may, under certain conditions, reach significant proportions when compared to conventional brand gasoline [6.9]. *Figure 6.5* shows a comparison of the characteristics of different fuels.

HC emissions measured under FTP-75 testing can be reduced by up to 40% if suitable fuels are used. Fuel composition also determines the hydrocarbons composition of the exhaust gas. Reactivity of the exhaust gas in terms of its ozone-formation potential (one of the features covered by Californian legislation) also is a function of fuel composition and may be reduced by up to 20% if suitable fuel formulations are used [6.41].

As reformulated gasoline has a different composition (e.g. lower content of olefins and aromatic substances, reduced volatility), more energy will be required to produce this type of fuel. This entails higher CO_2 emissions at the oil refinery, amounting to approx. 18% according to [6.38].

6.1.2 Diesel fuels

Fuel quality affects diesel engine emissions even more strongly than emissions of spark-ignition engines. HC, CO and particulate emissions are the main pollutants affected by the fuel quality. The effect on nitrogen oxides is negligible.

	Commercial gasoline	Reformed US gasoline
Density at 15°C	0,749	0,727
$\frac{RON + MON}{2}$	89,1	89,2
Reid steam pressure [bar]	0,65	0,60
Boiling range [°C]		
Start	30	31
10 %	50	51
20 %	66	62
30 %	82	73
50 %	105	94
70 %	126	115
80 %	148	132
90 %	177	136
95 %	196	187
End	221	214
Sulphur, weight [%]	0,04	0,026
Benzene, weight [%]	1,4	0,9
Paraffins, weight [%]	54,5	62,4
Naphthenes, weight %	3,0	3,0
Olefins, weight [%]	11,6	9,7
Aromatic compounds, weight [%]	30,9	19,4
MTBE, weight [%]	0	5,3

Fig. 6.5 Comparison of US petrol station gasoline and reformulated gasoline

Similar to gasoline for spark-ignition engines, diesel engine fuel is a mixture of hydrocarbons, although its boiling temperature is approx. 170 to 360 °C. Depending on diesel fuel composition and characteristics, the major differences focus on mixture formation and combustion and, hence, exhaust emissions. Major quality criteria are cetane rating, density, viscosity, boiling characteristics, aromatics content and sulphur content. With regard to environmental compatibility, the following requirements have to be met:

- Low density,
- Low content of aromatic compounds,
- Low sulphur content,
- High cetane rating.

To a certain degree, the above requirements contrast with the demands for engine power and fuel economy. This becomes evident when the relationship between parameters such as density, content of aromatic substances and ignition quality (readiness to ignite) is considered:

"High density = high content of aromatic substances = low ignition quality"

Important characteristics that are also related to the smoking tendency of hydrocarbons used for diesel engine fuels are shown in *Fig. 6.6* [6.2].

The effect of the content of aromatic substances on particulate emissions of a swirl-chamber diesel engine is explained in *Fig. 6.7*.

Since a high content of aromatic substances in the fuel reduces its ignition quality (readiness to ignite) but since at the same time particulate emissions are lower if the content of aromatic substances is reduced, particulate emissions also are a function of the cetane rating. *Figure 6.8* [6.6] shows the relative exhaust emissions as a function of the cetane rating for the HC, CO, NO_x and particulate components determined in FTP-75 tests.

6.1 Conventional fuels

	Ignitability	Properties at cold	Volumetric heating value	Density	Smoke tendency
n-paraffins	good	poor	low	low	low
i-paraffins	low	good	low	low	low
Olefins	low	good	low	low	average
Naphthenes	average	good	average	average	average
Aromatic compunds	poor	average	high	high	high

Fig. 6.6 Characteristics of hydrocarbons

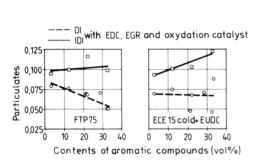

Fig. 6.7 Particulate emissions as a function of aromatic compounds for a swirl-chamber engine [Source: AVL]

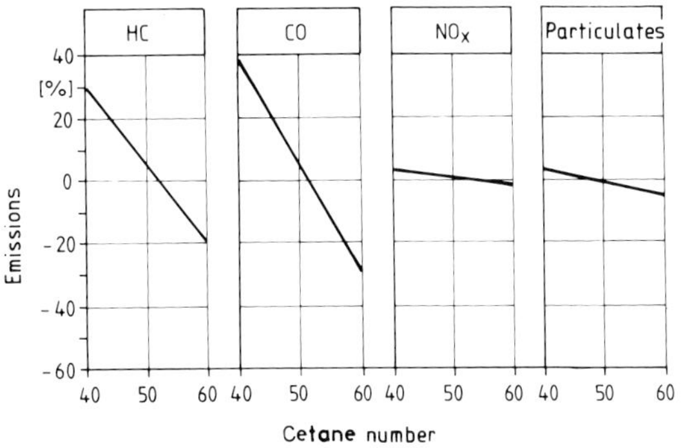

Fig. 6.8 Trends of relationship between exhaust emissions and cetane rating of a swirl-chamber engine

Figure 6.9 shows the particulate, soot and HC emissions as a function of engine load for two cetane ratings used on a supercharged direct-injection diesel engine [6.8].

With the exception of nitrogen oxides, the essential pollutant components can be modified by the cetane rating of the fuel.

The sulphur content of diesel fuel has a pronounced effect on particulate emissions. Sulphur produces both sulphates that contribute to particulate emissions and sulphur dioxide. Reducing the sulphur content of fuel is therefore a major issue today. The effect of such reductions, e.g. during ECE-15/04 testing, in terms of particulate emissions is illustrated in *Fig. 6.10*. The overall reduction is approx. 10%. Other sources [6.7] show results that indicate reductions of up to 30%. The kinetic characteristics of sulphate formation inside the engine appear to be related to the temperature level inside the engine, as in the case of NO_x formation. This means that higher engine loads accompanied by higher gas temperatures usually generate more SO_3, i.e. the mass of SO_3 per mass unit of sulphur increases along with the engine load.

Recent research results have shown that the sulphur content in the fuel should be reduced below a level of 0.02 weight percent if engine operation is taken into account. It remains to be investigated if a further reduction of the sulphur content is technologically and economically desirable. If the sulphur content is reduced to approx. 0.05 weight percent, as is already common practice in certain countries, CO_2 emissions during fuel production would increase by approx. 16% [6.39]. A further reduction of the sulphur content would cause the diesel fuel to lose its edge over spark-ignition engine gasoline in terms of CO_2 emissions that are 5% lower during production.

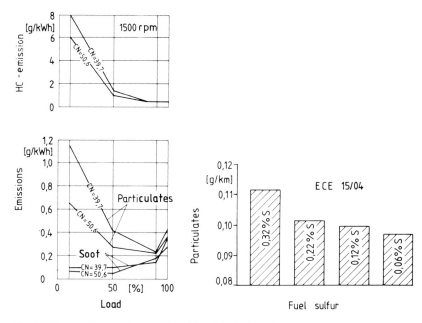

Fig. 6.9 Dependence of specific pollutants on cetane rating of a DI diesel engine [6.8]

Fig. 6.10 Effect of sulphur content on particulate emissions

The mean sulphur content levels of diesel fuels currently available in the Federal Republic of Germany vary between 0.08 and 0.19 weight percent [6.33]. The European limits to be applicable from October, 1996, are 0.05 weight percent.

6.1.2.1 Reformulated diesel fuel

In the same way as with gasoline, the amendment to the "Clean Air Act" enacted in the USA in 1990 also applies to diesel fuels [6.10]. A reformulated diesel fuel may have a low content of sulphur and aromatic substances [6.45] and should meet the following specifications:

- Improved readiness to ignite (e.g. CR >55),
- Reduced content of aromatic compounds (<10% by volume),
- Reduced density (e.g. 0.80 to 0.82 kg/l),
- Optimized addition of additives,
- 90% boiling point <300 °C.

This specification might yield the data shown in *Fig. 6.11*. This type of fuel will allow CO and HC emissions to be reduced by approx. 10% whereas particulate emissions can be reduced by up to 40%.

6.2 Alternative fuels

Alternative fuels have been analyzed and assessed according to a variety of criteria in the past [6.11]. Major criteria were and continue to be:

- Bottlenecks in the availability of crude oil,
- Reduction of dependence on imported crude oil.

6.2 Alternative fuels

	CEC RF-03-A84	Fuel with low contents of sulphur and aromatic compounds
Cetane rating	50,8	57,0
Density at 15 °C	0,839	0,809
S. weight %	0,24	< 0,005
Aromatic compounds vol%	36	< 3
Lower heating value	42,85	43,10
Boiling range		
Start	187	198
10 %	220	219
20 %	235	231
30 %	247	237
50 %	268	245
70 %	296	247
80 %	312	250
90 %	333	260
End	361	290

Fig. 6.11 Diesel fuel specifications

Another fact that has gained importance in recent years is the reduction of pollutant emissions by introducing alternative fuels. The following discussion shall focus on this latter item. In many cases, alternative fuels allow exhaust raw emissions to be reduced significantly, reducing the required expenditure for exhaust aftertreatment at the same time.

Exhaust aftertreatment systems, e.g. three-way catalytic converters for spark-ignition engines or oxidation catalysts for diesel engines, can also be used with alternative fuels. The most important alternative fuels suitable for engine use and covered by worldwide research and testing are:

- Alcohols (Methanol CH_3OH and ethanol C_2H_5OH),
- Petroleum gas (Propane C_3H_8 and butane C_4H_{10} in a mixture of approx. 50%/50%). This can be made available either as liquefied petroleum gas (LPG) or as compressed natural gas (CNG),
- Natural gas (Methane (CH_4) and other gases),
- Biogas (Methane and other gases, with CO_2 contents in biogas being up to 40%),
- Vegetable oils and vegetable oil methyl esters,
- Hydrogen (H_2).

With the exception of vegetable oils, all the above alternatives are suitable both for spark-ignition and for diesel engines. Vegetable oils are only suitable for diesel engines, whereas gaseous fuels are better suited to spark-ignition engines. Considering availability and structural constraints, hydrogen should definitely be considered a long-term alternative. Liquefied petroleum gas and biogas are only of regional importance. Natural gas is available in large quantities worldwide. Alcohols may be made available for use on a large-scale basis, and ethanol, for example, can be produced from plants and methanol.

The exhaust emission characteristics of alternative fuels not only depend on engine considerations but, to a decisive degree, also on the chemical and physical properties of the fuels. Combustion of alternative fuels made up of hydrocarbons produces the known

Property \ Fuel	Gasoline Premium	Diesel	Heavy fuel oil	Methanol	Ethanol	Vegetable oil (rape oil)	Liquefied petroleum gas	Methane	Biogas	Hydrogen
Density f=liquid kg/dm³ g=gaseous	f 0,73÷0,78	f 0,81÷0,85	f 0,95	f 0,79	f 0,79	f 0,92	f 0,54	g 0,72·10⁻³	g 1,20·10⁻³	g 0,09·10⁻³
Heating value MJ/kg	43,2	43	42,7	19,7	26,8	37,1	45,8	50	17,5	120
Mixture heating value (λ=1) kJ/m³	3750	3860	3660	3440	3475,0	3400	3725	3224	3200	3190
Ignition limit λ	0,4÷1,4	0,5÷1,35	0,5÷1,35	0,34÷2,0	0,3÷2,0	-	0,4÷1,7	0,7÷2,1	0,7÷2,3	0,5÷10,5
RON	98	-	-	111	106	-	100	105	-	-
MON	88	-	-	92	94	-	95	-	-	-
CN	-	45÷55	34÷44	-	-	40÷44	-	-	-	-
Molecular mass kg/kmol	97	191	198	32	46	882	51	16	27	2
Boiling temperature °C	30÷180	170÷360	175÷450	65	78	180÷360	-30	-162	-128	-253
Evaporation heat kJ/kg	420	300	-	1120	905	-	353	510	-	450
Steam pressure bar (n. Reid)	0,45÷0,9	-	-	0,37	0,21	-	-	-	-	-
m_L stoich kg_L/kg_{kr}	14,7	14,5	14,6	6,4	9	12,7	15,5	17,2	6,1	34
Composition kg_i/kg_{kr} h c o	0,15 0,85 -	0,137 0,863 -	0,15 0,85 -	0,125 0,375 0,5	0,13 0,52 0,35	0,12 0,77 0,11	0,177 0,823 -	0,25 0,75 -	0,09 0,44 0,47	1,0 0,0 0,0

Fig. 6.12 Properties of alternative and conventional fuels

pollutant spectrum, including CO_2. The higher the C content of the fuel, the higher are CO_2 emissions under equal conditions. One alternative that eliminates CO_2 emissions is the use of hydrogen.

Some key properties of alternative fuels compared to gasoline and diesel fuel are compiled in *Fig. 6.12*.

On the basis of the physical properties of fuels, it is possible to forecast the emission trends to be expected. As outlined in Chapter 2, high combustion temperatures produce increased nitrogen oxide emissions. When compared with other fuels combusted at the same air/fuel ratio, combustion of hydrogen reaches the highest adiabatic combustion temperature [6.25] and attains, hence, the highest NO_x level. All fuels having a high vaporization heat. e.g. alcohols, reach only lower combustion peak temperatures during the reaction and, hence, produce lower NO_x emissions. A wider span of the misfire limits of the fuel also has a positive effect on emission characteristics. The underlying reason is that pollutant concentrations directly depend on the air/fuel ratio (see Chapter 2). This facilitates implementation of quality management or of a combination of quality and quantity management.

6.2.1 Alternative fuels for spark-ignited engines

If we disregard vegetable oils for SI engines, the major fuel alternatives for spark-ignition engines are indicated above. Their behavior with regard to the major pollutant components is outlined below [6.12].

6.2 Alternative fuels

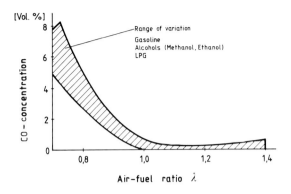

Fig. 6.13 CO concentrations during combustion of gasoline and alternative fuels [6.12]

- **Carbon monoxide**
  The scatterband of CO concentrations of some fuels is shown in *Fig. 6.13*. Significant differences are only evident in the stoichiometric range [6.12]. Of greater importance than the influence of fuel, however, are mixture formation and mixture distribution to the individual cylinders. The only advantage of the alternative fuels indicated in *Fig. 6.13* is their increased potential for lean-burn operation.

- **Unburned hydrocarbons**
  Figure 6.14 shows the relative HC emissions compared to gasoline [6.12]. When the misfire limits are approached, the scatterband of HC emissions of gasoline wides sharply. This is one area where the better lean-operation potential of alternative fuels offers

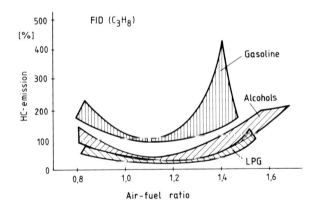

Fig. 6.14 HC emissions during combustion of gasoline and alternative fuels [6.12]

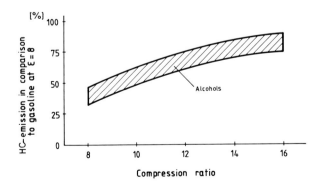

Fig. 6.15 HC emissions as a function of compression ratio [6.12]

particular benefits when compared to gasoline. The overall HC emissions caused by methanol, ethanol and LPG are lower than those of gasoline.

Increasing the compression ratio will also cause HC emissions to increase, even if alternative fuels are used (*Fig. 6.15*)[6.12]. This is explained both by the fact that exhaust gas temperatures are lower and post-reactions are therefore reduced, and by the impaired shape of the combustion chamber with an increasing surface-to-volume ratio and relatively large squish gaps.

- **Aldehydes**

When comparing alternative fuels in terms of HC emissions, the entire spectrum of organic components has to be taken into account. Aldehydes are one of the most important components of this spectrum, and aldehyde emissions are therefore already subject to regulation in the USA.

Figure 6.16 shows the relative aldehyde emissions vs the air/fuel ratio. It is evident that alcohols here fare worse than gasoline by a factor of 3 to 4 across major operational ranges [6.12].

- **Polycyclic aromatic substances**

Since the composition of fuel has a direct effect on emissions and since alcohol-based fuels do not contain polycyclic aromatic hydrocarbons, emissions in the exhaust also are one order of magnitude below those of gasoline engines.

- **Nitrogen oxides**

In line with the above statements, *Fig. 6.17* shows the relative NO_x emissions. Alcohols offer particular benefits as far as NO_x emissions are concerned since combus-

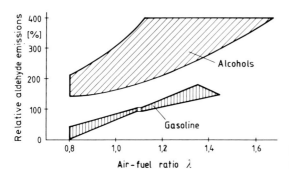

Fig. 6.16 Relative aldehyde emissions of alcohols and gasoline [6.12]

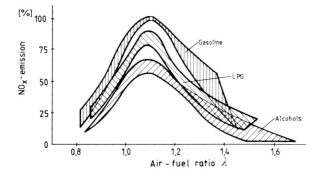

Fig. 6.17 Relative NO_x emissions of alternative fuels [6.12]

6.2 Alternative fuels

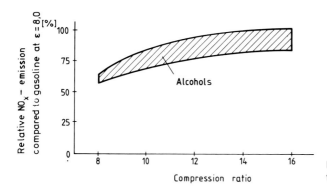

Fig. 6.18 Relative NO$_x$ emissions as a function of compression ratio [6.12]

tion temperatures are relatively low and operation at lean air-fuel mixture settings is possible.

Since alcohols are highly knock-resistant, compression ratios can be increased beyond the level of gasoline engines. Part of the above benefits will then be lost, however, as both the compression final temperature and, hence, the combustion temperature will be higher.

In spite of this, considerable advantages remain even if the full potential of higher compression ratios is utilized [6.18] (*Fig. 6.18*).

NO$_x$ emissions are, among other features, dependent on the H/C ratio of the fuel. If we disregard pure hydrogen combustion, the following rule applies: The higher this ratio, the lower are nitrogen oxide emissions during combustion [6.13].

- **Ozone**

  As explained above, the ozone content in the lower layers of the air is a decisive factor in assessing air quality. If the relative ozone formation of gasoline-operated passenger cars is assumed to be 100%, the ozone-formation potential could be reduced by approx. 70% thanks to the lower CO and HC content if all vehicles were operated with methanol. This is one of the main reasons why the Californian Air Resources Board (CARB) demands a more widespread use of methanol-operated vehicles.

6.2.1.1 Alcohols (methanol and ethanol)

The advantages of using alcohol fuels are lower NO$_x$ and HC emissions and increased suitability for lean settings (*Fig. 6.19*). Methanol also offers benefits in terms of CO$_2$ emissions since its specific energy consumption is lower than that of gasoline (*Fig. 6.20*).

The use of alcohol fuels offers further advantages:

- Accelerated natural decomposition,
- Ozone formation in the lower air layers is reduced thanks to differences in exhaust composition (e.g. HC spectrum),
- Lower evaporation rate.

One problem linked to the use of alcohols is increased aldehyde formation.

Using methanol alone (without incorporating suitable exhaust aftertreatment systems) will not allow future emission control standards to be met.

Another option would be to use methanol-gasoline fuels, e.g. M90. This is a mixture consisting of 90% methanol and 10% gasoline. The results obtainable with this fuel mixture in conjunction with catalytic converters are shown in *Fig. 6.21* [6.14].

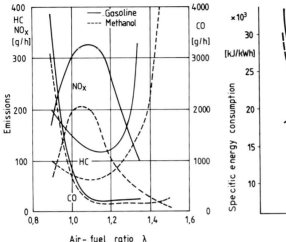

Fig. 6.19 Exhaust emissions of gasoline and methanol [6.47]

Fig. 6.20 Comparative diagram of specific energy consumption [6.48]

Emissions and fuel consum.		M90	Gasoline	Change in % *)
CO	[g/kWh]	40,24	51,52	78
HC	[g/kWh]	1,15	5,26	22
NOx	[g/kWh]	8,30	12,61	66
HCHO	[g/kWh]	0,56	0,37	151
be	[g/kWh]	667	392	170
Efficiency	[ ]	0,248	0,217	114

Part load n= 2500 rpm, we= 0,25 kJ/dm³
*) Gasoline operation =100%

Fig. 6.21 Exhaust emissions and fuel consumption of medium-sized vehicles according to FTP-75 testing [6.47]

Since this mixture does not contain the highly volatile components, special measures or devices are required to facilitate cold starting (e.g. fuel or air preheaters).

6.2.1.2 Gases

Liquefied petroleum gas (LPG) is composed mainly of propane and butane. These components allow the exhaust carbon dioxide content to be reduced significantly as their hydrogen content is far higher. As LPG also offers better lean-burn operation characteristics than gasoline, HC values can be reduced considerably as well. In the case of nitrogen oxides, lower emissions are only achieved at lambda settings above 1.4, as shown in *Fig. 6.22*. Lower λ values will produce a sharp increase of NO_x emissions. *Figure 6.22* also underlines that emissions of unburned hydrocarbons increase in the same way as with gasoline operation at increasing excess air ratios (although at far higher λ values) [6.34].

Present-day emission standards cannot be met with LPG operation unless a catalytic converter is fitted. The usable λ range approaching lean mixture settings is wider than in the case of gasoline operation. As an added bonus, knock tendency is reduced. As a result, engine compression ratios can be raised. This will reduce exhaust temperatures, and the catalytic converter will therefore have to be designed for the prevailing temperature levels. The main requirement therefore is to improve the light-off performance of the catalytic converter.

6.2 Alternative fuels

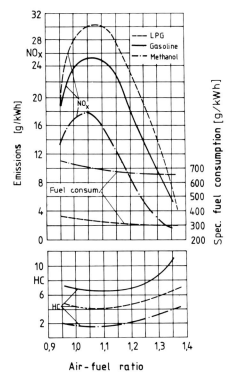

Fig. 6.22 Comparison of specific emissions of different fuels [6.14]

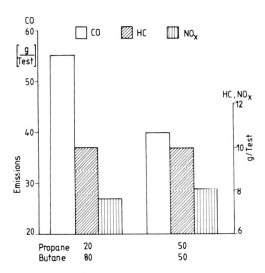

Fig. 6.23 Effect of LPG composition on emission characteristics [6.15]

As with other fuels, the composition of LPG has a profound effect on emission characteristics. *Figure 6.23* shows the ECE test results for different contents of propane and a butane [6.15].

6.2.1.3 Hydrogen

The use of hydrogen fuel for internal-combustion engines is an alternative that will probably gain considerable long-term importance. The advantage of hydrogen is that a closed matter or energy circuit is formed if solar energy is used to produce hydrogen. As an added bonus, hydrogen operation offers a slightly higher efficiency [6.25]. Hydrogen produced from water in a process involving solar energy reacts to form water during combustion. The environmental characteristics of engines operated with hydrogen should be considered extremely positive when compared to engines using conventional fuels [6.11] [6.16] [6.17] [6.18].

Due to the fact that hydrogen does not contain any hydrocarbons, exhaust emissions theoretically do not contain any carbon dioxide, carbon monoxide and unburned hydrocarbons. In practice, HC emissions do occur but their magnitude in terms of environmental pollution is negligible. These emissions are primarily caused by the combustion of lubricating oil. The content of oils in the combustion process of current engines is approx. 0.01 to 0.05 l/100 km and can be reduced even further. NO_x is the component that is present in increased quantities due to excessive combustion peak temperatures. *Figure 6.24* shows the basic characteristics of pollutant concentrations vs the air/fuel ratio. Combustion of hydrogen also leads to increased hydrogen and hydrogen vapor emissions.

In contrast with other fuels (e.g. gasoline, methanol), hydrogen can be ignited within a very wide limit. The range available for engine operation is approx. $\lambda = 1$ to 7. Due to its properties, hydrogen may be used as fuel in two different ways:

- Hydrogen is used to substitute fuels containing hydrocarbons.
- Hydrogen is used as a fuel for specific operating ranges or as an additive to improve combustion and to facilitate ignition during multi-fuel operation.

Thanks to the extremely wide misfire limits of hydrogen, the engine can also be operated with quality management in order to improve fuel economy. This means that, similar to diesel engine practices, a load increase is achieved by enriching the mixture. This is possible both with pure H_2 operation and with H_2 mix operation.

The effect of H_2 operation on refueling, storage, infrastructure, hydrogen production etc. will not be discussed here any further.

a) Pure hydrogen operation: Hydrogen may be added to the combustion air in two ways:

- Into the intake manifold,
- Into the combustion chamber.

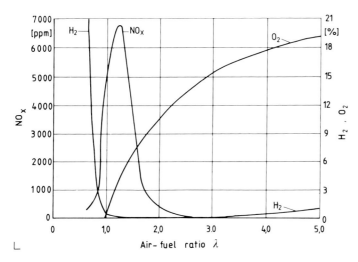

Fig. 6.24 Exhaust pollutant concentrations as a function of air/fuel ratio of engines operated with H_2 [6.20]

6.2 Alternative fuels

The first alternative is a case of external mixture formation, the second is called internal mixture formation. Similar to other gaseous fuels, hydrogen also has an edge over liquid fuels as far as mixture formation is concerned. One particular benefit is the extremely high diffusion capacity of hydrogen since this contributes greatly to ensuring an equal supply of fuel to all cylinders [6.19].

If hydrogen is added in a liquid state and is transformed into a gaseous state during mixture formation, the mixture cools down sharply, thus reducing NO_x formation and, hence, NO_x emissions. The process temperature, on the other hand, is somewhat higher. *Figure 6.24* shows the exhaust composition (without water) of an engine operated with hydrogen as a function of the air/fuel ratio [6.20].

In the range below stoichiometric settings, unburned fuel (H_2) emissions reach appreciable levels. In contrast with conventionally operated engines, this range is not too significant in the case of external mixture formation since quality management can be implemented with this design.

The important range between $\lambda = 1$ and approx. $\lambda = 5$ shows that H_2 emissions will also occur, and the effect of this type of emissions on the environment will have to be investigated in large-scale usage. In contrast with engines operated with hydrocarbons, the shape of the combustion chamber and the surface-to-volume ratio is of lesser importance for H_2 emissions. This is explained by the fact that hydrogen is able to sustain combustion even in narrow gaps. The extremely long and flat shape of the combustion chamber of rotary engines actually lends itself better to the use of hydrogen [6.36].

Depending on the degree of excess air present during combustion, nitrogen oxide emissions are expected to be comparatively low [6.16].

Other important measures to reduce NO_X emissions with hydrogen operation are exhaust recirculation and water injection. At the same time, these measures also effectively help prevent backfiring that may occur during hydrogen operation with external mixture formation. Both methods have been adopted for real experimental engines.

The physical properties of hydrogen promote high exhaust recirculation rates (up to 40%). This allows NO_x emissions in CVS tests to be reduced below 0.04 gpm [6.21].

Odorous emissions are not known to have been produced by hydrogen-operated engines, yet higher NH_3 emissions should be taken into account.

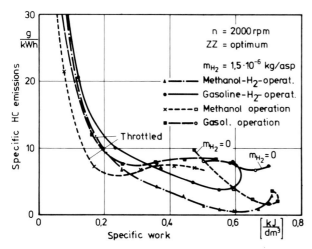

Fig. 6.25 Specific HC emissions during mixed operation vs methanol and gasoline operaton [6.22]

Fig. 6.26 Results of ECE and FTP 75 tests with a hydrogen-gasoline mixture

ECE - Test	
Component	g/Test
CO	8,70
HC	3,00
NOx	0,40
CVS - Test	
Component	g/mile
CO	0,45
HC	1,57
NOx	0,51

b) Hydrogen mix operation: Thanks to its wide misfire limits, hydrogen also is an ideal partner to be mixed with methanol or gasoline in order to reduce exhaust emissions [6.25].

On spark-ignition engines, may be accomplished using quality management. In this case, the engine is operated with the base fuel at full throttle, hydrogen is added to the fuel in the part-throttle range and pure hydrogen is used at idle.

The effects of this type of multi-fuel concept on vehicle weight, refueling, infrastructure etc. are not considered here. To illustrate this type of mixed operation, *Fig. 6.25* shows the HC emissions [6.22].

This diagram compares the emissions of methanol-hydrogen and gasoline-hydrogen operation at a constant hydrogen mass per cycle to the emissions obtained when operating the engine with the pure basic fuel only. The curves show that operating the engine in a fuel mix mode allows relatively low HC emissions to be achieved in the medium and high load ranges. This benefit in terms of reduced emissions, however, is not sufficient to enable this design to meet the legal emission standards without the use of an exhaust aftertreatment system (catalytic converters). This is highlighted in *Fig. 6.26*, using ECE and FTP-75 tests as an example. These tests were carried out on a vehicle fitted with an engine operated with a mixture of hydrogen and gasoline [6.23].

6.2.2 Alternative fuels for compression ignition engines

6.2.2.1 Alcohols (methanol and ethanol)

The use of alcohol fuels is an attractive option not only for spark-ignition engines but also for compression-ignition (diesel) engines [6.18] [6.24]. The primary advantage of this type of fuel is that it allows exhaust emissions to be reduced.

The increased vaporization heat of methanol and the resulting temperature reduction of the cylinder charge are two assets that facilitate reduction of NO_x emissions. Exhaust recirculation is also possible with methanol operation. An added advantage is that the oxygen contained in methanol provides for virtually soot-free combustion and, hence, allows particulate emissions to be reduced to the low level typical of spark-ignition engines. A drawback of both spark-ignition and compression-ignition engines operated with methanol, however, is that aldehyde emissions are higher. Emissions of these partially oxidised hydrocarbons may be reduced by using oxidation catalysts, however.

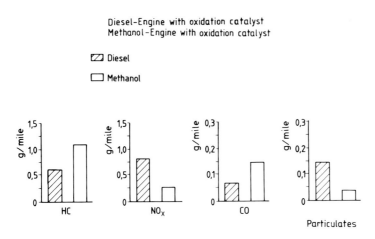

Fig. 6.27 Emission levels of methanol and diesel fuel operation [6.24]

6.2 Alternative fuels

Figure 6.27 compares the characteristics of methanol and diesel fuel operation according to FTP-75.

The slightly higher HC and CO emissions of engines operated with methanol may be explained by the effect of the high vaporization heat level on the combustion process [6.24]. In the lower part-load range, in particular, combustion is slowed down at high λ values.

Alcohol operation of diesel engines requires improved self-ignition characteristics of the alcohols used. Ignition accelerators, ignition jet methods using a pilot quantity of diesel fuel, extremely high combustion ratios and the use of spark plugs or glow plugs are just some of the methods that may be implemented to accomplish this goal. Additionally, the injection system must be adjusted for the increased fuel quantities (low heating value).

6.2.2.2 Vegetable oils

In addition to alcohols produced from biomass (e.g. ethanol), vegetable oils are another fuel that can be produced using raw materials from sustainable sources. This is referred to as a closed circuit. Using vegetable oils as fuels only makes sense, however, if the respective biofuels can be produced with far less conventional energy than is contained in the final biofuel product (cf. *Fig. 6.28*) [6.26]. The energy balance must yield definite positive results.

In order to achieve a closed energy circuit, it should also be considered that, in addition to the cultivating surfaces available, a significant number of additional areas will have to be

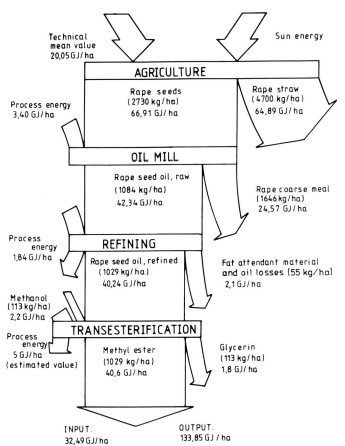

Fig. 6.28 Energy balance of growing raw materials [6.26]

activated, thus entailing a variety of related problems, e.g. monoculture, excessive use of fertilizer, contamination of groundwater etc.

A wide variety of vegetable oils may be considered for potential use as fuels in diesel engines. In addition to soybean oil, peanut oil, cocoanut oil, palm oil etc., rape seed oil is discussed as a particularly important raw material. Pure, untreated oils can cause malfunctions in diesel engines as their viscosity is fairly high and, hence, atomization is poor. This results in incomplete combustion and carbon deposit formation on the injection nozzles. The polymerization tendency of these oils accelerates resin and soot formation. Suitable measures must therefore be introduced to prevent such processes from occurring when vegetable oils are used as fuels. Transesterification of the oils is one method of overcoming such problems. This transesterification process is accomplished by adding alcohols to the oil. As a result, glycerine-water and ester-alcohol mixtures are formed as soon as a certain temperature level is attained and when a catalytic converter is present.

After separating and distilling the excess alcohol, this process yields the ester usable or combustion in engines. This process is suitable e.g. for producing rape seed methyl ester (RME) suitable as "diesel fuel" from raw rape seed oil. RME will be investigated here in greater detail as it is typical for a variety of vegetable oil derivatives.

6.2.2.3 Rape seed methyl ester (RME)

Engine operation with RME offers special advantages in terms of soot emissions [6.27] [6.28] [6.31]. At full throttle and across the entire rpm range, the soot numbers of engines using RME are approx. two Bosch units below the values of engines operated with diesel fuel (*Fig. 6.29*). This applies both to swirl-chamber engines and to direct-injection engines and may be explained by the oxyen contained in RME.

In the lower load range, specific fuel consumption figures of engines operated with RME are somewhat higher than with diesel fuel. This is due to the boiling characteristics [6.29] since the boiling temperature of RME is far higher (*Fig. 6.30*) and therefore affects mixture preparation in the lower load range adversely. An additional characteristic is that, due to the higher viscosity of RME, small injection quantities cause more atomization problems and therefore also tend to impair mixture preparation.

Figure 6.1 shows the results measured in FTP-75 tests for a swirl-chamber diesel engine operated wih RME. Apart from lower HC and CO figures and roughly equal particulate emissions, aldehyde and NO_x emissions are higher. Other investigations [6.29] have

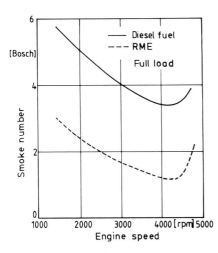

Fig. 6.29 Soot numbers of engines operated with diesel fuel and rape seed methyl ester [6.28]

6.2 Alternative fuels 135

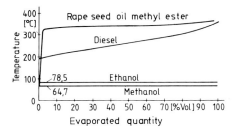

Fig. 6.30 Boiling curves of different fuels

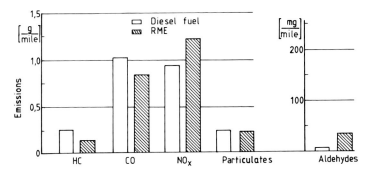

Fig. 6.31 Emissions of rape seed methyl ester and diesel fuel according to FTP-75 [6.28]

yielded somewhat lower NO_x emissions. The oxidisable HC and CO components and the soluble particulate contents as well as odorous substances can be reduced to a very high degree by using oxidation catalysts. Since the advantages that RME offers over diesel fuel in terms of exhaust emissions are not significant and since criteria such as overall ecological balance and availability have to be taken into account, RME does not appear to be the fuel that will help solve future exhaust emission problems. In terms of environmental protection, however, RME has the advantage of being able to decompose itself in the ground to a degree of 98% within 20 days. This makes RME particularly suitable for specific market niches such as agriculture, motorboats, runway maintenance vehicles etc.

Recent research has shown that biologically produced fuels are no suitable alternative if overall ecological criteria such as environmental pollution caused by production, distribution etc. are taken into account. Manufacturing costs (without taking fuel tax into account) currently still are approx. 4 times as high as the cost of conventional diesel fuel [6.42].

6.2.2.4 Other fuels

Figure 6.32 shows the effect of different fuels on lean-burn operation and emission characteristics.

Figure 6.33 attempts to evaluate the major systems and alternatives including electrical drive systems [6.30]. The column representing electrical drives also shows the effect on exhaust emissions if electrical power is generated by nuclear energy. The lack of emissions of conventional pollutant components makes this drive system a very attractive alternative, although power generation using nuclear energy produces other types of emissions that are not taken into account within the scope of this discussion.

When studying the object-related emissions e.g. of vehicle engines, the level of the pollutant emissions often depends on the applicable system limits. The emissions caused by

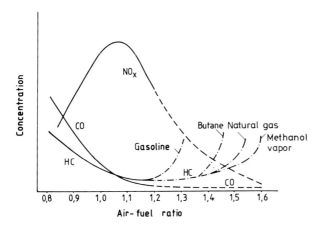

Fig. 6.32 Basic pollutant concentration curve of different fuels [6.46]

	Gasoline fuel w/o cat.conv.	Gasoline fuel w cat.conv.	Diesel	LPG [4]	Natural gas	Hydrogen	Bio-ethanol	Methanol	Electrical	
Suitability	0	0	0	−	−	−	0	0/−	− −	
Availability	0	0/−	0	−	0	(+++)[1]	− −	− −	− − [4]	
Economy	0	0	+	+/−	−	− − −	− −	−	− −	
Infrastructure	0	0/−	0	− −	− − −	− − −	− − − [3]	− − − [3]	− −	
									Carbon	Nuclear
CO	0	+ +	+ +	+	+	+++	+	+	+	+++
HC	0	+ +	+ +	+	+	+++	+ [5]	+ [6]	+	+++
NOx	0	+ +	+	−	−	+/−	0	0	0	+++
Particulates	0	+	− − −	+	+	+++	+	+	−	+++
CO2	0	0	+	+	+ +	+++	+ +	0/− − [2]	− −	+++
Fuel toxicity	0	0	+	+++	+++	+++	+++	+++	+++	

+ slightly better   − slightly worse
++ noticeably better   − − noticeably worse
+++ very much better   − − − very much worse
0 = equal

[1] In water
[2] Due to manufacturing process
[3] When abandoning use of current fuel
[4] For electricity : 0
[5] Aldehydes : −
[6] Dual system without catalytic converter

Fig. 6.33 Evaluation of alternative fuel systems

stored processes that are generated e.g. during production, transport, storage etc. are disregarded in most cases. If these emissions are taken into account, the overall situation often changes dramatically, as highlighted in *Fig. 6.34* [6.40].

Such emissions are not necessarily emitted in regions that have to cope with heavy basic pollution, i.e. global HC emissions do not change significantly if an electrical vehicle is used instead of a diesel vehicle. Locally, for example in metropolitan areas, air quality can indeed be improved by such measures.

6.3 Effects of fuel additives

Fuel additives serve two distinct purposes. One purpose is to improve combustion and pollutant emissions, another is to ensure reduced wear and limit deposit formation during

6.3 Effects of fuel additives

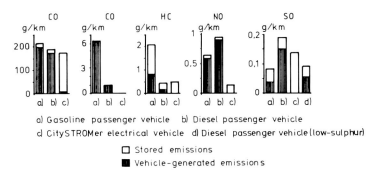

a) Gasoline passenger vehicle b) Diesel passenger vehicle
c) CitySTROMer electrical vehicle d) Diesel passenger vehicle (low-sulphur)

☐ Stored emissions
■ Vehicle-generated emissions

Fig. 6.34 Stored emissions of fuel, diesel and electrical vehicles [6.40]

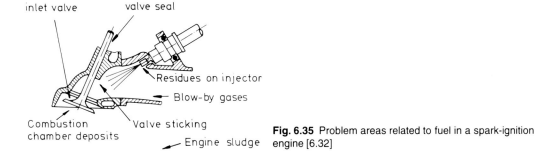

Fig. 6.35 Problem areas related to fuel in a spark-ignition engine [6.32]

the engine life cycle of several hundreds of thousands of miles. *Figure 6.35* shows some important fuel-relevant problem areas in a spark-ignition engine [6.32].

Whereas the solution of a technological problem tended to dominate in the past, the effects on the environment are given increasing priority today. One particular trait of additive manufacturers and users is that the chemical composition of the various additives is largely kept secret. Discussions must therefore be limited to reporting the effects of such additives.

6.3.1 Additives for gasoline engines

The most important additives that have a direct effect on combustion are "anti-knock products" such as tetra ethyl lead and tetra methyl lead. Following the introduction of unleaded fuel, however, these products have been superseded since the heavy metals they contain are both toxic and unsuitable for use with catalytic converters (catalyst contamination). For the same reasons, other additives containing metals that are designed to improve combustion (combustion enhancers) are no longer used in Germany today.

Of greater importance today are additives that have a positive effect on long-term engine characteristics and that compensate differences of fuel composition that may have been caused by the use of different crude oils and different refining processes.

Two examples are presented to illustrate the positive effect of multifunctional additives. *Figure 6.36* shows how exhaust emissions are affected adversely by deposit formation and residues inside the combustion chamber and on the inlet valves. Additives will help avoid these problems.

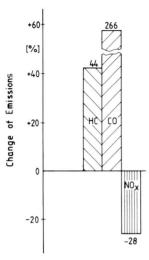

Fig. 6.36 Effect of deposit formation on pollutant emissions [6.32]

If no additives are used, deposits will accumulate on the valve heads that act like a sponge during cold starting and cause the air-fuel mixture to become leaner. Drivability will suffer eventually. At full throttle, the deposits reduce the cylinder charge and thus cause power losses.

6.3.2 Additives for diesel engines

In contrast with external mixture formation in spark-ignition engines, mixture formation in diesel engines occurs within a far shorter time frame. The effect of additives concentrates on improving combustion, reducing deposit formation during combustion of the fuel and on keeping the injectors clean.

Whereas it is attempted to prevent fuel self-ignition in spark-ignition engines by selecting a high octane rating, self-ignition of diesel engines, on the other hand, must be enhanced by using a high cetane number. As only a very short time is available for this process, there is always a risk of incomplete combustion and, hence, excessive hydrocarbon and soot emissions. Excessively high combustion speeds at high temperatures, on the other hand, result in higher nitrogen oxide emissions.

Three groups of additives are used for diesel engines:

- *Cetane number enhancers:* Increasing the cetane number by using cetane number enhancers means that ignition lag is reduced, i.e. ignition is advanced. Ignition lag is shown schematically in *Fig. 6.37*. Extensive research has proven that increased additive concentrations help to reduce HC, CO and particulate emissions by a significant degree without causing NO_x emissions to increase. When the cetane number was increased by ten units, HC emissions dropped by up to 80%, CO emissions were up to 50% lower and particulate emissions were reduced by up to 30%. Combustion noise is also reduced considerably.
- *Combustion enhancers:* Whereas the effect of cetane number enhancers concentrates mainly on the start of combustion, combustion enhancers act directly on the combustion process. This effect is achieved both in clean and in coked engines (in contrast with detergents that first have to accomplish their cleaning effect.) Unburned hydrocarbons and particulates are reduced by a significant degree whereas CO and NO_x emissions

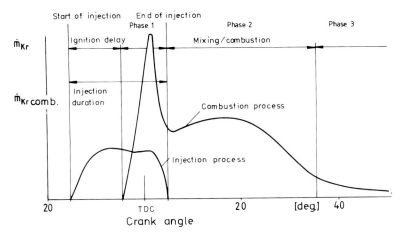

Fig. 6.37 Schematic diagram of ignition lag

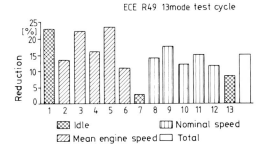

Fig. 6.38 Effect of combustion enhancers on particulate emissions

remain unchanged. When considering particulate emissions in the 13 individual phases of the ECE-49 test, it is noticed that particulates are reduced to a somewhat greater degree than at nominal speed (*Fig. 6.38*) [6.32]. This may probably be explained by the fact that increased particulate formation, as observed during this test, means that the reduction potential is also increased.

- *Multifunctional diesel additive:* This additive pack includes components for corrosion protection, demulsification and reducing the foaming tendency. Oxidation inhibitors and detergents are used to keep the injectors clean. The effect of this additive pack was proven on two taxi vehicles with swirl-chamber engines that had covered a mileage of 80,000 miles without replacement of the injectors. After a 600-mile cleaning run with the additive pack, HC emissions had dropped by 45% and CO and particulates were more than 10% lower without causing any significant increase of NO_x emissions. The positive effect of this type of additive has since been proven on direct-injection diesel engines as well.

6.4 Effects of lubricants

Present-day high-performance engines (both spark-ignition and compression ignition designs) will meet the imposed requirements only if high-performance engine oils are used. Following the introduction, among others, of multi-valve technology, supercharging, catalytic converters, variable valve timing, acoustic capsules as well as through extended oil change intervals, reduced oil capacities and reduced oil consumption, the demands on lubricants have undergone drastic changes.

Reduced oil consumption is of particular benefit to catalytic converters as this reduces the risk of catalyst contamination and extends catalyst life [6.49].

Reasons for the reduction of oil capacities are reduced weight and faster warming of the oil after cold starts.

The current state of the art of oil change intervals is 9,000 to 12,000 miles, and extended intervals following the same trend have also been specified for diesel engines.

High oil temperatures have to be overcome by using oil coolers since the active agents in engine oils undergo chemical changes above temperatures of 150 °C. Above a level of approx. 160 °C, the mineral-based basic oil components start to vaporize. Oxidation of engine oil will also increase exponentially with temperature, with the reaction speed doubling for every 10 to 15 °C.

Special attention must be paid to cold temperature characteristics as well as to high temperature stability, wear protection, engine cleanliness and to compatibility with engine seal materials. The effect on the catalytic converter and environmental compatibility are also of importance.

Favorable low-temperature operating characteristics can be achieved by using synthetic all-year multigrade oils of 5W-40 or 5W-50 viscosities. These oils are sufficiently thin for cold starting (5W) and thus speed up oil supply to the lubricating points. Even when the engine is hot, viscosity still is sufficient (40 or 50 grade). This type of oil allows fuel consumption to be reduced by 3 to 4% as friction losses are reduced significantly in the highly dynamic range. Mineral-based 15W-40 multigrade oil is used as a reference oil for

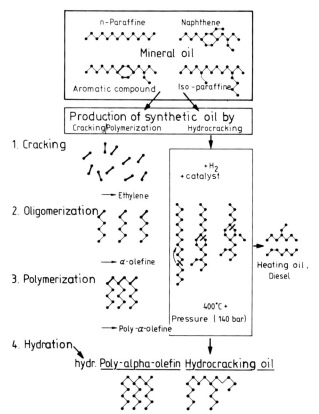

Fig. 6.39 Production processes of synthetic oils [6.44]

6.4 Effects of lubricants

this comparison [6.43]. Synthetic oils are particularly suitable as base oils since they are highly resistant to ageing [6.44]. 0W-30 or 0W-40 light-running oils are currently under development.

Two base oils are commonly used: Poly-alpha-olefins (PAO) and synthetic esters (sometimes used in conjunction with hydrocracking oils). *Figure 6.39* shows the synthesis production process. Using catalytic control, molecules that provide the desired characteristics and size are produced chemically from mineral oils.

For environmental reasons, these new engine oils have lower evaporative losses and phosphorus content of 0.1%. This prevents contamination of the catalytic converter, thus helping to extend its service life. Optimizing the additive packs also allows the sulphatic ash content to be reduced, therefore keeping oil contamination low (i.e. wear is reduced).

In addition, the content of chlorine (an inactive agent produced during additive production) has also been reduced. This facilitates recycling of used oils.

The additives used all contribute directly or indirectly to reducing exhaust emissions and fuel consumption. They are:

- Detergents (Multifunctional agents with cleaning and cleanliness additives, with corrosion protection and basic reserves for long oil intervals).
- Dispersants (Agents to prevent redeposition of contamination, combined with highly molecular components to prevent sludge formation).
- Anti-wear additives (Wear protection agents both for short-distance motoring and for high-temperature operation).
- Anti-oxidants (Agents to prevent oxidation of the engine oil that remain active even at extended oil change intervals).
- Viscosity index (V.I.) enhancers with high shear stability.

7 Problems with CO_2 emissions

7.1 CO_2 emissions and their causes

When considering the problem of CO_2 emissions, a distinction must be made between the sources of pollution. For the purpose of this study, only anthropogenic emissions, i.e. emissions from technology systems created by man, will be considered. These emissions essentially are the result of energy conversion processes. Approx. 750 million tons of carbon dioxide were emitted in Germany (without the former East German territory), corresponding to 3.6% of the worldwide anthropogenic CO_2 production. Road traffic accounts for approx. 20% of the overall output. As shown in *Fig. 7.1*, this percentage tends to increase even further if primary energy consumption is taken as a measure for CO_2 emissions [7.1] [7.2].

7.2 CO_2 emissions from motor vehicles

If the consumed fuel quantity and the resulting CO_2 emissions are considered from a stoichiometric point of view, using oxidation of a fuel equivalent substance, e.g. isooctane, as an example, the below relationship is obtained. Nitrogen reactions are not taken into account here since they are of no relevance to the below basic relationships.

$$C_8H_{18} + 12.5\,O_2 \Leftrightarrow 8\,CO_2 + 9\,H_2O$$
$$1\,kg\,C_8H_{18} + 3.509\,kg\,O_2 \Leftrightarrow 3.088\,kg\,CO_2 + 1.421\,kg\,H_2O$$

In the case of iso-octane (the assumed fuel equivalent), combustion of 1 kg of fuel at stoichiometric conditions therefore produces approx. 3 kgs of CO_2. This establishes a direct relationship between CO_2 emissions and fuel consumption of a motor vehicle across the combustion process.

According to [7.3], the below formula can be used to determine CO_2 emissions of passenger cars fitted with spark-ignition and diesel engines. This formula takes the measured fuel consumption and pollutant emissions into account.

$$^mCO_2 = 0.85 \cdot m_{fu} - 0.429 \cdot CO - 0.866 \cdot HC/0.273$$

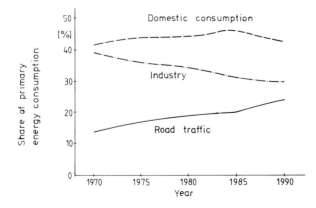

Fig. 7.1 Share of road traffic in consumption of primary energy

7.2 CO₂ emissions from motor vehicles

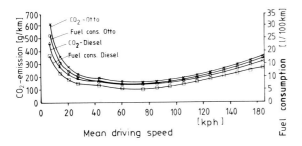

Fig. 7.2 CO$_2$ emissions and fuel consumption as a Function of vehicle speed [7.2]

Where: m_{fu} = fuel mass, and CO and HC = emission coefficients of carbon dioxide and unburned hydrocarbons.

On the basis of the above relationships, *Fig. 7.2* shows the CO$_2$ emissions as a function of vehicle speed along with the corresponding fuel consumption figures. This diagram assumes that oxidation to yield CO$_2$ occurs to the maximum extent possible.

Studies [7.12] describe research carried out to determine the CO$_2$ emission characteristics of spark-ignition and diesel engine vehicles as a function of vehicle mass in FTP-75 tests. As shown in *Fig. 7.3*, the results show an advantage of approx. 20% for pre-chamber type diesel engines. According to [7.13], this advantage in terms of CO$_2$ emissions will remain unchanged in the future.

If the dependence of carbon dioxide concentrations on the air/fuel ratio is examined, the curve shown in *Fig. 7.4* is obtained [7.4]. The maximum is reached at a stoichiometric air/fuel ratio. CO$_2$ concentrations are lower both in the lean and the rich setting. It should be taken into account, however, that the CO emissions released into the atmosphere that are particularly high in the "rich" range will eventually be retransformed into CO$_2$. If the engine is operated at lean settings with lower CO$_2$ emissions, engine output will drop accordingly.

Since reducing the fuel consumption is the only way of reducing CO$_2$ emissions if fossil fuels are used, diesel engines, and direct-injection diesel engines in particular, offer considerable advantages in this respect. *Figure 7.5* shows the respective figures obtained for diesel and spark ignition engines on the basis of U.S. city tests.

Figure 7.6 gives an overview of the efforts required to reduce pollutant emissions with different drive systems and the relevant emission standards. The effort required to achive zero emissions (including zero CO$_2$ emissions) is extremely high.

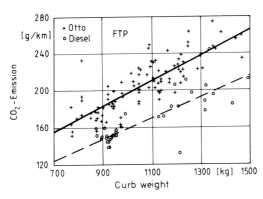

Fig. 7.3 CO$_2$ emissions of passenger vehicles according to FTP testing [Source: German Federal Environmental Agency]

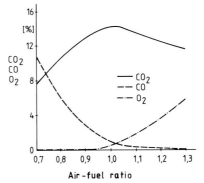

Fig. 7.4 CO$_2$, CO, and O$_2$ concentrations as a function of air/fuel ratio [7.4]

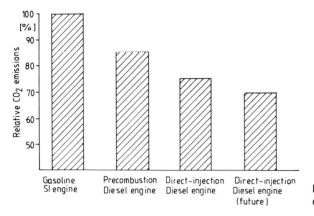

Fig. 7.5 CO_2 emissions of different engine concepts

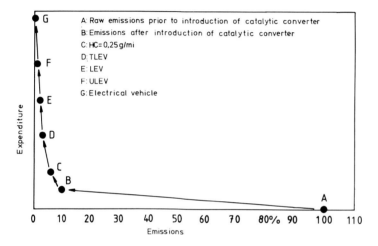

Fig. 7.6 Emissions of different drive systems [7.15]

7.3 CO_2 emissions and fuels

Apart from the combustion process, both the fuel and its production also have a considerable effect on CO_2 emission levels.

If we compare different drive concepts in terms of CO_2 emission characteristics, the entire energy chain will have to be taken into account. This interrelationship is shown in *Fig. 7.7* for the three major primary energy sources, i.e. crude oil, natural gas and coal [7.5] [7.6].

A CO_2 emission coefficient of one is attributed to diesel fuel produced from crude oil that offers typical diesel-engine combustion efficiencies.

Only methane and methanol produced from natural gas have emission coefficients that are superior to the coefficients achievable with conventional diesel fuels.

Fuels produced from coal are particularly unsuitable with regard to CO_2 generation.

If only the areas of generation, transport, processing and distribution are taken into account in this assessment of CO_2 emissions, diesel fuel has an edge of approx. 5% [7.14] over gasoline. Most of this difference is duel to the different production processes used to produce diesel and gasoline fuels (approx. 100 kg/ton for diesel fuels compared to approx. 230 kg/ton for gasoline) [7.13].

7.3 CO_2 emissions and fuels

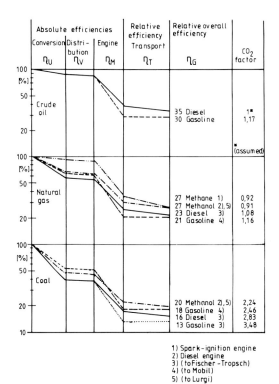

Fig. 7.7 Energy chain and CO_2 coefficient [7.4]

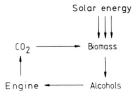

Fig. 7.8 Closed CO_2 circuit [7.7]

Fuels produced from biomass are part of a closed circuit and therefore cause less pollution of the environment. In fact, this circuit is not really closed since considerable amounts of energy are required to produce biomass. Additional problems of increased soil and water pollution caused by fertilizers and plant protective agents as well as by monocultures will also have to be addressed. New crop areas would also be required or, alternatively, the yield per unit area would have to be increased (see Chapter 6.2). This circuit process is illustrated in the schematic diagram in *Fig. 7.8*.

If we compare the emissions of conventional and alternative fuels as an equivalent to CO_2, the trend shown in *Fig. 7.9* is established [7.9] [7.10]. The emissions relative to the mileage are also indicated. If the problem of the greenhouse effect becomes more acute in the future – and there is nothing to indicate that this will not be the case – and if the problems that are invariably linked to more widespread use of nuclear energy are taken into account, only two alternatives remain in addition to cutting down energy use drastically. One alternative is to produce fuels from biomass, which means that the abovementioned problems will have to be overcome, and the other is to use solar energy for the production of hydrogen and of electrical energy.

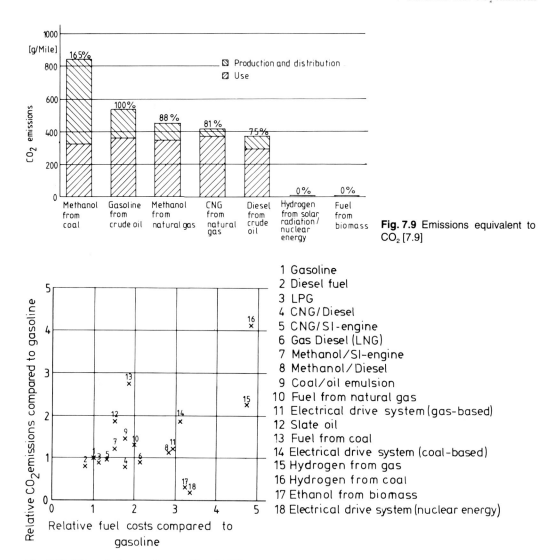

Fig. 7.9 Emissions equivalent to CO_2 [7.9]

Fig. 7.10 CO_2 emissions and expenditure [7.10]

Apart from technological and environmental requirements, the applicability of alternative fuel concepts is also subject to cost constraints. *Figure 7.10* shows the relative CO_2 emissions in terms of the relative fuel costs [7.12]. Those alternatives that may contribute to reducing CO_2 emissions significantly still entail costs today that amount to three to five times the price of gasoline [7.11]. Alternative No. 18 shown in *Fig. 7.10* should be considered as being fairly critical in terms of environmental compatibility.

8 Laws regulating the emission of pollutants and maximum fuel consumption of combustion engines (as of 1992)

The increased air pollution that is a major source of concern in virtually all highly industrialized nations with a high car/population ratio has led to the introduction of measures to limit pollutant emissions [8.2] [8.9] [8.14] [8.15]. To ensure reproducibility and comparability, the emitted pollutant quantities are determined according to specified test methods. One drawback is that test methods, classifications and emission standards often differ from one country to another, i.e. direct comparisons are only possible to a limited extent or even not at all. In addition, the large variety of test methods adopted for emission measurement entails unnecessarily high development costs. The excessive proliferation of numerous different international standards, measuring methods and test cycles will have to be brought to an end, not least of all in order to minimize costs.

8.1 Testing procedures

The basic aim was to create testing cycles that have to be driven by the vehicle in order to determine the emitted pollutants in a reproducible manner. Subsequent assessment according to predefined criteria will yield emission figures that are then compared to the legal standards. The engines will have to be designed in such a manner that emissions remain below the respective legal limits.

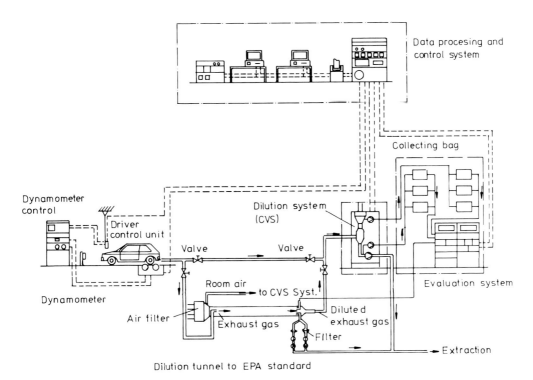

Fig. 8.1 Schematic diagram of test dynamometer with exhaust emissions analysis system

The vehicle is driven through a predefined test cycle on a roller-type dynamometer. According to the time sequence of the preset vehicle speed, the engine is operated in specific load-rpm ranges. The dynamometer features a brake dynamometer setup that allows the operating loads to be adjusted. Driving resistances such as air and rolling resistance and friction are determined, for example, by coasting tests and are then taken into account by entering corresponding characteristic curves. Depending on the vehicle weight, the roller-type dynamometer is fitted with rotating masses that are used to simulate vehicle inertia during acceleration. The test cycles, i.e. the time-speed profile, are usually stored in a driver control panel. *Figure 8.1* shows a schematic diagram of a system that is designed for driving the defined test cycles.

In addition to the US test, a European test cycle modified by the EEC Environmental Council in 1989 is currently specified for the Federal Republic of Germany and for the EEC territory. This cycle (*Fig. 8.2*) is subdivided into an urban driving cycle (Part 1) and into an extra-urban driving cycle (Part 2) [8.2]. Part 1 is completed four times, starting with a cold start on a roller-type dynamometer. The extra-urban test cycle specifies test speeds of up to 120 km/h (75 mph) and is completed once. This new driving cycle ensures that the test matches real-world driving conditions more accurately.

The city cycle includes extremely high idling time fractions (31%) and coasting operation periods. The average driving speed is only 19 km/h (12 mph). Although the ECE cycle was designed to represent the driving characteristics in actual road traffic, it does not meet this goal. It should also be noticed that only a small part of the engine map is covered by Part 1 of the European test cycle, as shown in *Fig. 8.3* for an actual case.

This means that only emissions within this operating range of the engine are subject to legal standards. All other sections of the map could be designed according to criteria other

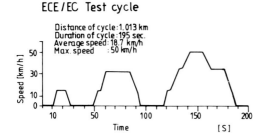

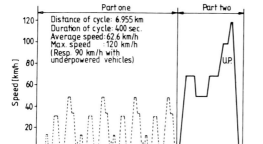

Fig. 8.2 European Test Cycle [8.13]

8.1 Testing procedures

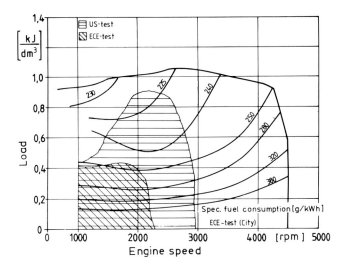

Fig. 8.3 Engine map areas covered by test cycles

than minimizing pollutant emissions. Obviously this would contradict the intentions of emission control legislation.

The first and, hence, oldest driving program is the "California cycle" introduced in California in 1966. It is no longer used today as it was replaced by the US City Test in the early 70s. FTP-75 testing (Federal Test Procedure) is the current state of the art in the USA and in other countries. Its sequence is shown in *Fig. 8.4* [8.13].

The test sequence is subdivided into four sections. Exhaust emissions of the individual sections are collected individually in sampling bags and are then processed. The individual test sections comprise:

- Transient phase with a duration of 505 sec and an assessment coefficient of 0.43,
- Stabilized phase up to 1,371 sec with an assessment coefficient of 1.0,
- Stop phase of 10 min,
- Driving a transient phase in a repetitive mode (hot starting), with an assessment coefficient of 0.57.

Before starting the exhaust emissions test, the requirements for preparation of the vehicle must be met. The most important requirements are:

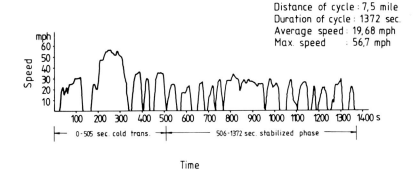

Fig. 8.4 FTP-75 cycle

- The vehicle must be conditioned for 12 to 36 h at a temperature between 20 °C and 30 °C.
- A flywheel mass according to the operational mass of the vehicle is specified.

The representation of the US city cycle within the engine characteristic map (*Fig. 8.3*) shows that, unlike the urban European test, this test covers a far higher segment of collective load and rpm conditions.

8.2 Measurement methods and instruments

8.2.1 Emission measurement methods

Standardized measurement methods are used for both the ECE and the FTP-75 tests. The CVS (Constant Volume Sampling) method is used for this process [8.3].

When driving the vehicle in this test cycle, the exhaust gas is diluted with ambient air and is collected in sampling bags. A blower feeds the thinned exhaust at a constant displacement.

The exhaust mass is determined from the blower speed and the temperature of the gas mixture [8.9]. The mean volumetric air-to-exhaust ratio must be no less than 8:1 so that condensation can be avoided. A metering pump extracts the samples into the bag. This process is carried out according to a predetermined sequence. At the same time, part of the flow of ambient air required for dilution is sampled in bags. This is necessary to allow the ambient air to be checked for pollution. The bag contents are then analyzed. The pollutant mass of the individual components is computed from the mass of the supplied gas and the pollutant concentration in the exhaust. The pollutant mass determined in this process may be referenced to the respective test or the test duration. The emission values can then be indicated in grams of pollutant per test or in grams of pollutant per mile.

Figure 8.5 shows an excerpt of the concentration curve (showing the extended starting phase) in FTP-75. The pollutant concentrations are indicated both upstream and downstream of the catalytic converter. This diagram shows that most of the pollutant emissions are produced during the first two to three minutes while the operating temperature of the catalytic converter still is too low to enable sufficient conversion of pollutants. This is one aspect that requires further development to improve emissions.

A number of legal requirements have to be observed during gas and particulate sampling. The most important requirements are:

- Exhaust gas must be diluted with sufficient ambient air in order to avoid formation of condensation water in the sampling and measuring systems,
- The total exhaust and dilution air volume must be measured exactly,
- A part flow of diluted exhaust gas and diluted air must be sampled continuously for analysis purposes,
- The sample concentrations are corrected in accordance with the content of gaseous air pollutants contained in the ambient air.

A number of other requirements and legal regulations specify standards for volume measurements, sampling methods, sampling system, computation of emission mass etc. These as well as other standards are specified in the German Official Federal Gazette, Part 1 [8.10].

Diesel engine particulate measurements as specified e.g. in the U.S.A. are performed in a dilution tunnel. The design of such a system is shown in the schematic diagram in *Fig. 8.6*.

8.2 Measurement methods and instruments

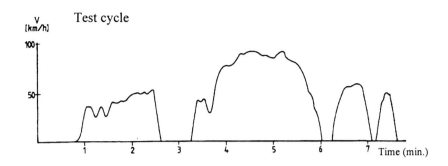

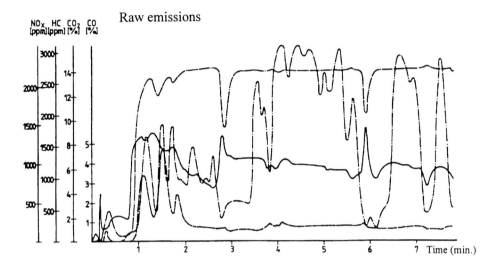

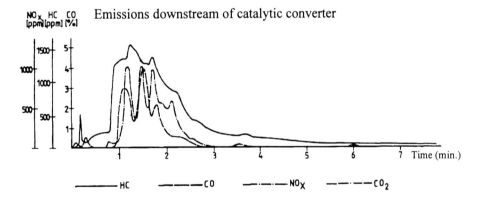

Fig. 8.5 Example of pollutant concentrations emitted during the CVS test

The diluted exhaust is extracted from the dilution tunnel and is then ducted across filters. The degree of filter charging is an indicator of the particulate content in the exhaust. A CVS system located downstream of the tunnel is used to fill the exhaust sampling bags for analysis of gaseous components. The gas sampling system must be designed in such a way that the mean volumetric CO_2, CO, HC and NO_x concentrations can be measured continuously [8.5].

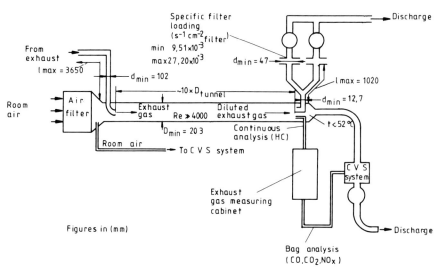

Fig. 8.6 Dilution tunnel [8.5] [8.40]

8.2.2 Emission measuring instruments

The exhaust gas sampled according to the specified procedures must be analyzed. In this context, it is important to measure not only the content of components subject to limitations but also the O_2 and CO_2 components used as reference substances. The concentrations of gaseous components are measured with gas analyzers while particulates are measured with special traps. Selective measuring processes tailored to suit the component to be measured are used for gaseous components. Solid matter, e.g. particulates, is detected gravimetrically on the basis of filter loading or of the discoloration of the filter paper.

To ensure that correct measurement technology is applied that also meets the legally specified boundary conditions, various physical measurement principles have been introduced. Usually the below analyzers are used to measure the emission components that account for the major part of emissions contained in the diluted exhaust [8.6]:

- NDIR (non-dispersive infrared) measuring instruments for carbon monoxide and carbon dioxide,
- Flame ionization detectors (FID) for hydrocarbons,
- Chemical luminescence detectors (CLD) for nitrogen oxides,
- Particulate measuring instrument (dilution tunnel and trap).

The most important measuring instruments and devices currently used to analyze spark-ignition and diesel engine exhaust emissions are:

- NDIR photometer: This allows CO and CO_2 emissions to be measured. Basically this device is also suitable for measuring hydrocarbon components. These, however, are usually measured with a FID. The photometer detects the absorption of infrared radiation caused by the measuring gas. This measuring process is based on the property of multi-atom gases of absorbing radiation in the infrared spectral range (2.5 to 12 µm). The wavelengths of the absorption bands indicate the type of gas while the degree of absorption is an indicator of the concentration of the component measured. Depending on the type of gas and the measuring range, two different measurement principles are used. Either the substances can be compared, or an interference-filter correlation is run. The design of an infrared photometer is shown schematically in *Fig. 8.7*.

8.2 Measurement methods and instruments

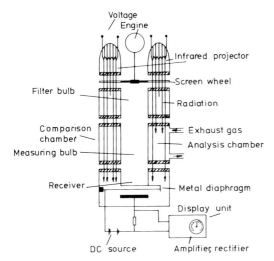

Fig. 8.7 Measurement principle of the NDIR photometer [8.39]

The wide-band radiation emitted by infrared radiation units passes an orifice wheel to enter the analysis or comparison chamber. The radiation emitted from these chambers is captured by detectors. The difference between both detector signals created by the timed sequence of reference and measuring filters is an indicator of the concentration of the component in the measurement gas.

- FID (Flame ionization detector): The concentrations of unburned hydrocarbons in the exhaust are usually measured with a flame ionization detector. Due to the wide spectrum of unburned hydrocarbons, the analyzer must be calibrated with a typical component present in the exhaust. A very clear ion flow is produced if an electrical field is applied to a pure hydrogen flame. The hydrogen flame is fed by a supply of hydrogen and air that is free from hydrocarbons. If the measuring gas adds hydrocarbons to this flame, the ion flow increases. This increase is proportional to the number of hydrocarbon atoms. Since the FID measures the sum of all organically bound hydrocarbon atoms in the hydrocarbons, calibration is performed with a component that is representative of the exhaust, e.g. C_6H_{14} or C_3H_8. The magnitude of the ion flow is an indicator of the HC concentration in the exhaust. *Figure 8.8* shows the flow diagram and design of this measuring instrument [8.4].
- CLD (Chemiluminescence detector): The NO or NO_x concentration in the engine exhaust is usually determined using the chemiluminescence effect. This is based on the reaction of nitrogen monoxide with ozone according to the following equation:

$$NO + O_3 \Leftrightarrow NO_2 + O_2 + h \cdot v$$

The light quantum emitted in this process is detected by a photomultiplier and is converted into an electrical signal that is proportional to the NO concentration.
- Oxygen analyzer (Residual oxygen contents in the exhaust): The measurement principle is based on the paramagnetic property of oxygen. Permanent magnets in the measuring chamber that is passed by the measuring gas create a strongly non-homogeneous magnetic field. If oxygen is present in the measuring gas, the oxygen molecules are drawn into the non-homogeneous magnetic field due to their paramagnetic behavior. The highly non-homogeneous character of the magnetic field creates oxygen particulate pressures of different magnitudes. These particulate pressures are higher at locations showing a high field strength than at locations of low field strength. The pressure differential between the

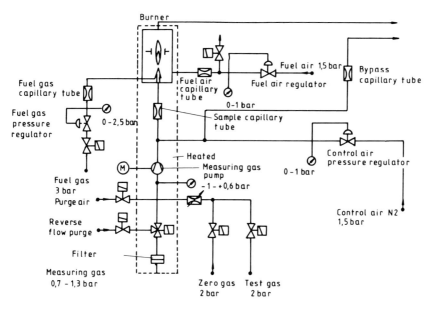

Fig. 8.8 Flow chart and circuit design of a FID [8.4]

partial pressure of the measuring gas and the partial pressure of a reference gas indicates the oxygen concentration in the exhaust gas.
- Smoke analyzer: The soot or smoke number is measured by a diesel smoke analyzer. This analyzer measures the soot content in the exhaust of diesel engines in a photoelectrical process using the filter paper method. During this measurement, an extraction sensor extracts a specified exhaust flow from the engine and draws it in across a special filter paper having a predefined surface. The soot particulates in the exhaust cause the filter paper to assume a greyish color. This grey value is detected by a reflection photometer and is correlated with a smoke scale rating. A 100% reflection corresponds to a smoke number of zero while a 0% reflection corresponds to a smoke number of ten. *Figure 8.9* shows the basic design of such a system [8.4].
- Particulate measuring instrument: *Figure 8.10* shows a schematic diagram of a particulate measuring system used for stationary engine operation.

The main element of this system is a dilution tunnel that is used to compute the particulate emissions using the exhaust mass, the particulate mass accumulated on the

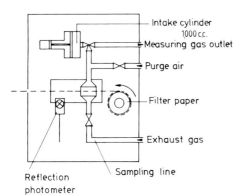

Fig. 8.9 Basic design of a diesel smoke meter [8.1]

8.2 Measurement methods and instruments

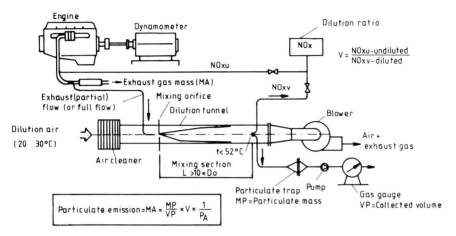

Fig. 8.10 Schematic diagram of a particulate measuring system [8.8]

Propane in air	Hexane in N_2	CO in N_2		CO_2 in N_2	NO in N_2
ppm	ppm	ppm	%	%	ppm
from	from	from	from	from	from
8	50	7,5	0,3	0,08	5
in 18 steps up to	in 8 steps up to	in 22 steps up to	in 11 steps up to	in 18 steps up to	in 15 steps up to
18000	1000	2500	10,0	16,0	9000

Fig. 8.11 Overview of required test gas mixtures [8.6]

particulate trap, the gas volume fed across the particulate trap and the NO_x dilution ratio [8.8].

Dilution tunnels are designed either as full or part-flow systems. In the case of part-flow systems, only a partial flow of the engine exhaust gas is removed and diluted continuously.

All the measurement devices described above, with the exception of the diesel smoke and particulate analyzers, have to cope with the drawback that different calibrating gases must be available in a variety of concentrations. *Figure 8.11* presents an overview of the test gas mixtures that are required in order to cover the entire range of engine exhaust measurements [8.6].

This entails considerable financial and logistic expenditure for the test gas supply system. A recent newly-developed exhaust analyzer avoids this drawback. With this instrument, CO_2 is the only type of calibration gas required [8.7].

Sampling is another important aspect of the engine exhaust analyzing process. Sampling has a decisive effect on the accuracy of the measurements.

A heated sampling device is particularly important for sampling unburned hydrocarbon emissions. This type of sampling allows the exhaust gas to be ducted from the sampling point to the measuring point across an electrically heated line. This approach also eliminates system failure that may be caused by condensation in the sampling line.

8.3 Pollutant limits and maximum fuel consumption values

8.3.1 USA—49 states

Although the population density of the United States is comparatively low when compared to Europe and Japan, air pollution has been a major problem in the USA for many years. For one thing, the USA are a worldwide leader in energy consumption and, hence, in the use of fossil energy sources [8.12], and in addition, air pollution is particularly critical in certain metropolitan areas such as Los Angeles (*Fig. 8.12*). Due to the special topographic and climatic conditions, severe air pollution caused by automotive exhaust emissions has been a problem in this region since the early 60s [8.11].

Starting in 1968, and followed by the "Clean Air Act" of 1970, emission standards were specified for passenger cars in the USA. From 1977 to 1981, more stringent standards were gradually introduced. Upon enactment of the 1981 standards, the introduction of a three-way catalytic converter became unavoidable in the USA.

In 1982, additional limits were introduced for diesel passenger cars in order to limit particulate emissions. This standard was made more stringent in 1987. Due to the special situation in California, this state enforced even more restrictive measures than the other 49 states. U.S. laws also specify that durability of the emission control systems must be proven for a period of 5 years or 50,000 miles. The 1972 U.S. standards were based on a driving cycle that was designed to reproduce traffic conditions in Los Angeles (LA-4 Cycle or FTP-72). In 1975, this driving program was extended to include additional test stages and is since referred to as FTP-75 [8.14] [8.9].

To reduce energy consumption, the Energy Policy and Conservation Act was passed in 1975. The aim of this Act was to limit fuel consumption of passenger vehicles to 27.5 miles

Fig. 8.12 Main areas of pollution in the United States of America

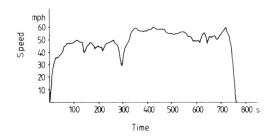

Fig. 8.13 U.S. Highway test cycle

8.3 Pollutant limits and maximum fuel consumption values

Passenger vehicles with spark-ignition or Diesel engines </= 12 persons			
Test method	Pollutant	Dim.	Limits
US 75 FTP	HC	g/mile	0.41
	CO		3.4
	NO$_x$		1.0
	Particulate		0.20
SHED	Evaporation	g/test	2.0

Fig. 8.14 Current U.S. 49-States limits

per U.S. gallon by 1985. This figure refers to the corporate average mileage of each manufacturer and is calculated from an average fuel consumption figure based on sales of all vehicles of a manufacturer. These fuel consumption standards known as CAFE standards (Corporate Average Fuel Economy) are calculated on the basis of an exhaust test (City Test) and of a fuel consumption test (Highway Test) (*Fig. 8.13*).

An additional U.S. specification is intended to limit evaporative emissions of spark-ignition engined vehicles. These emissions are measured in the SHED test (Sealed Housing for Evaporative Emissions Determination) [8.9] [8.16].

The U.S. test method shown in *Fig. 8.15* features three distinct sections, i.e. test descriptions, conditions for testing factory-new vehicles and conditions for monitoring vehicles in use [8.14].

Figure 8.14 shows the current limits [8.13] [8.17].

When fuel consumption is determined in accordance with the CAFE standard, it should be noticed that in countries that have adopted U.S. emission measurement methods, fuel consumption is calculated according to the "Carbon Balance Method" on the basis of exhaust composition. The overview below shows the basic relationship (*Fig. 8.16*, p. 159).

8.3.2 USA – California

The extremely high air pollution in the California region is due both to the geographic location of this state (South Coast Air Basin) that causes air exchange to occur only relatively slowly, and to the climatic conditions in this region. The number of sunshine days per year is very high, i.e. photochemical reactions lead to increased ozone production. The federal ozone standard, for example, was exceeded on 176 days in this region in 1988 [8.11]. The Californian emission limits valid up till 1992, are indicated in *Fig. 8.17* (p. 160). Both California and the other states of the USA are subject to the FTP 75 emission test procedure and to the fuel consumption limits based on the CAFE standard [8.11].

8.3.3 Japan

A large share of air pollution in Japan is caused by road traffic. By 1980, the vehicle population had increased to more than ten times the 1960 level. Within a very short span of time, however, from 1973 to 1977, vehicle emissions were reduced to a very low level.

At the same time, emission standards have been made increasingly stringent, with separate emission standards being enacted for diesel-engined passenger vehicles from 1981 [8.12] [8.18].

Along with the market introduction of the first vehicles fitted with catalytic converters, unleaded fuel was offered in Japan from 1975, and from this time, the use of unleaded fuel was specified virtually throughout.

158   8 Laws regulating the emission of pollutants and maximum fuel consumption

FTP-75	Continuous driving program according to the "City (Emissions Test) Cycle" (also referred to as LA-4 cycle).
	Test sections: 1. Cold start and cold transient phase (0 to 505 s) 2. Stabilized phase (506 to 1372 s) – 10-minute stop (Cooling fan off, hood closed) 3. Hot start and hot transient phase (1373 to 187 s). Speed sequence as during cold transient phase.
	Exhaust emissions and/or ambient air of each of the three test sections are sampled in three separate bag pairs and are then analyzed, multiplied with different evaluation coefficients and added to the overall result. The evaluation coefficients are: 0.43 for cold transient phase, 1.0 for stabilized phase, 0.57 for hot transient phase. Transmission operation as for FTP-72.
Evaporation Test (California)	As of Model Year 1995, a new, more comprehensive evaporation measurement procedure will be introduced gradually for Californian vehicles (for phase-in, refer to "California" exhaust table). The sequence of the new evaporation test is as follows: 1. Loading the carbon canister with butane in accordance with a specified procedure. – 2. After completing the FTP-75 exhaust test, a "Running Loss Test" is run (to determine evaporative emissions caused during vehicle operation). The RL test consists of three LA-4 cycles (= FTP-72) run on a dynamometer at an ambient temperature of 40.6 °C. Emissions are determined either with SHED closed (usual measuring method) or with the test system open (point source measurement at special locations of the vehicle fuel system). – 3. One-hour Hot Soak test at 40.6 °C. – 4. Three consecutive 24-hour heating tests in SHED at cyclic temperature change in 24-hour cycle of 18.3 → 40.6 → 18.3 °C (= Diurnal Test).
	Limits (HC or OMHCE in the case of methanol vehicles): Running Loss Test: 0.05 g/m Hot Soak Test + Diurnal Test: 2.0 g/test
	1. Testing of new vehicles *1.1 Routine production tests*
Inspection Test (California)	Each vehicle has to pass a functional test of all emission-related components and systems according to the manufacturer's specifications during production. This test specification is subject to a formal approval procedure. If systems or components are modified, the test steps will have to be revised accordingly.
Quality Audit Test (California)	The QAT (Quality Audit Test) is an sampling-type emissions test. The results of this test can be used to determine emission conformity of the production model. For this purpose, a random sampling of 2% of the production batch is subjected to an emissions test according to a certification procedure (FTP-75). The QAT results have to be reported to the authorities via quarterly reports. If the vehicles fail one single emissions test according to the applicable emission standards (deterioration factors (DFs) from 50,000/100,000-mile certification test runs have to be taken into account), a repeat test may be carried out. Both the re-test and any repairs or readjustments must be recorded in the test log. The QAT is considered to be complied with if the arithmetical mean of the sample (after applying the applicable DFs) complies with the valid emission standards. "Functional Tests" and "Quality Audit Tests" are specified in Title 13 CCR, Section 2061, "Cal. Assembly Line Test Procedures".
	1.2 Non-routine production testing
SEA (49 states)	The "Selective Enforcement Audit" (SEA) program (CFR, Subpart G, Section 86.601-86.615) is a sampling emissions test run by the authority at the manufacturer's premises in order to verify compliance with applicable emission standards. A predefined number of vehicles (sample lot) is subjected to an FTP-75 test and must meet a defined minimum quality level (Acceptable Quality Level = AQL; here, 60% of the sample lot meets the standard). The evaluation of the test result (number of vehicles from the sample that do not meet the standard) with regard to the AQL is performed on the basis of Sample Inspection Criteria Tables. Requirements for routine production checks in the manufacturer's plant or for report duty do not exist.
Surveillance Testing (California)	Official test of a vehicle manufacturer's new vehicles according to Title 13, CCR, Section 2150. Mercedes-Benz supplies vehicles from the "Vehicle Preparation Center" in Los Angeles for this purpose. The test covers functional checks and emissions tests according to FTP-75.
Dealer Surveillance (California)	Official surveillance test of new vehicles at the dealer's premises according to Title 13, CCR, Section 2151. In this case, new vehicles are submitted to an abbreviated test that may consist e.g. of an idling test, ignition timing check etc.
	2. Monitoring of vehicles in use *2.1 Non-routine monitoring by authorities*
In-Use Test (49 states)	Emission sampling test (FTP-75 test procedure) of vehicles with mileages below 50,000/75,000 miles within the EPA "Surveillance Test" programs.
In-Use Test (California)	Emission sampling test (FTP-75 test procedure, evaporation test) of vehicles in use (under 50,000/75,000 miles, depending on the type of certification of the vehicles concerned) according to Title 13 CCR, Sections 2136 to 2140.
	3.2 Continuous indirect vehicle monitoring by the manufacturer based on processing of warranty claims
Defect Reporting (California)	As of Model Year 1990, in-use vehicles will be subject to mandatory reporting of claims or defects of defined emission control system components and/or systems. This type of mandatory reporting covers a period of 5/10 years or 50,000/100,000 miles, depending on the warranty duration of the component or assembly (Title 13, CCR). The reporting procedure is based on three reporting steps with increasing detailing of the report statements, i.e. "Emission Warranty Information Report (EWIR)" (Section 2144), "Field Information Report (FIR)" (Section 2145), and "Emission Information Report (EIR)" (Section 2146). These reports allow information or complaints, fault rates, fault analyses and emission results to be forwarded to the authorities. The FIR and EIR reports are used by the authorities for decision finding that may be used to enforce manufacturer recall campaigns (Section 2136–2140).
Defect Reporting (49 states)	As of Model Year 1972, in-use vehicles are subject to mandatory reporting of defects or damages of defined emission components/systems. This type of reporting is mandatory if at least 25 identical emission-related components of one Model Year suffer from a defect. This mandatory reporting covers a period of 5 years after the end of the Model Year. In addition to a listing of the relevant components, the report includes a description of the defect, an evaluation of its effect on emissions, and indications of remedial action recommended by the manufacturer. The report is used by the authorities for decision finding linked to recall obligations imposed on the manufacturer.

Fig. 8.15 Description of U.S. test procedures (excerpt), translation from [8.13]

8.3 Pollutant limits and maximum fuel consumption values

$$\text{Diesel fuel consumption [mpg]} = \frac{K}{0.866 + (HC) + 0.429 - (CO) + 0.273 \times (CO_2)}$$

Gasoline fuel consumption [mpg]

$$= \frac{(5174 - 10^4 \times CWF \times SG) \, *)}{[(CWF \times HC) + (0.429 \times CO) + (0.273 \times CO_2)] \times [(0.6 \times SG \times NHV) + 5471]}$$

CWF: Hydrocarbon content ASTM D 3343 ⎫
SG: Fuel density ASTM D 1298 (g/ml) ⎬ Replace with actual
NHV: Net heating value ASTM D 3338 (BTU/LB) ⎭ values of fuel analysis

USA: The fuel economy (FE) figures calculated along with the emissions test certification values by the US EPA are calculated from the "City Test" and the "Highway Test" according to the following formula:

$$FE = \frac{1}{0 \times 55/CFE + 0 \times 45/HFE} \text{[mpg]}$$

with:

CFE = FE from City Test (FTP 75) [mpg]
HFE = FE from Highway Test [mpg]

The sales-weighted total of the vehicles marketed by a manufacturer in the USA must meet the below FE specifications based on the above formula:

Model year	1978	1979	1980	1981	1982	1983	1984	1985	1986	1987	1988	1989	1990	1991
mpg	18	19	20	22	24	26	27	27,5	26,0	26,0	26,0	26,5	27,5	27,5
l/100 km	13,06	12,38	11,76	10,69	9,80	9,05	8,71	8,55	9,05	9,05	9,05	8,88	8,55	8,55

Conversions: (mpg = miles per gallon)

$$\frac{USA}{ECE} \frac{235.215}{US \text{ mpg}} = \frac{l}{100 \text{ km}}$$

$$\frac{GB}{ECE} \frac{282.5}{Imp. \text{ mpg}} = \frac{l}{100 \text{ km}}$$

$$\frac{USA}{J} 0.4255 = \text{mpg US} = \frac{km}{l}$$

$$\frac{ECE}{J} \frac{100}{l/100 \text{ km}} = \frac{km}{l}$$

Fig. 8.16 Calculation of fuel consumption [8.13]

In order to counteract the increasing fuel consumption levels, the Japan Ministry of Transport enacted the Energy Saving Act in 1979. This program required the automotive industry to reduce the fuel consumption of passenger vehicles by 12% from 1978 to 1985 (This requirement was extended to imported vehicles as of 1988) [8.18].

Japanese emission regulations are relatively complicated. Passenger car tests are based on five test procedures and two test cycles, i.e.:

- 10-mode test cycle: This cycle represents urban driving conditions. After prior conditioning of the vehicle, the cycle starts with a hot start. Maximum vehicle speed is 40 km/h. This test cycle is also used in order to determine fuel consumption.
- 11-mode test cycle: This cycle starts with a cold start, with vehicle speeds of up to 60 km/h being reached. It was introduced in 1975 in order to complement the 10-mode cycle and is only applicable to passenger cars with spark-ignition engines.
- 3-mode test: This test applies only to diesel-engined passenger cars and is used to determine smoke and soot emissions.
- Idle exhaust test: This test is run within the preconditioning phase of the 10-mode test and is used additionally to check HC and CO emissions at idle.

1980–1992 Gasoline and Diesel passenger cars

Primary standards

Manufacturers certifying new vehicles to the following standards must demonstrate compliance at 50,000 miles. In 1981 and 1982, manufacturers had the choice of certifying new vehicles to option 1 or option 2 listed below. In 1989, manufacturers must certify no more than 50% of their vehicles to the 0.7 g/m option. In 1990–93, manufacturers must certify no more than 10% of the previous years production to the 0.7 g/m NO_x standards. Those vehicles certified to the optional 0.7 g/m NO_x standard are subject to a 7-year/75,000 mile recall for selected emission control parts. See additional standards listings for other emission requirements.

Year	Hydrocarbons non-methane	total	CO	NO_x	Notes
1980	0.39 g/m	0.41 g/m	9.0 g/m	1.0 g/m	
1981	—	0.41	3.4	1.0	option 1
1982	0.39	0.41	7.0	0.4	option 1
1981	0.39	0.41	7.0	0.7	option 2
1982	0.39	0.41	7.0	0.7	option 2
1983–88	0.39	0.41	7.0	0.4	
	0.39	0.41	7.0	0.7	optional
1989–92	0.39	0.41	7.0	0.4	
	0.39	0.41	7.0	0.7	optional

Optional 100,000 mile Gasoline and Diesel passenger car standards

Manufacturers have the option of certifying new vehicles to the following 50,000/100,000 mile standards. Manufacturers must demonstrate compliance with both the 50,000 and 100,000 mile standards for hydrocarbons and carbon monoxide and a 100,000 mile NO_x standard. For the 1989 and later model year, only diesel passenger cars may certify to these standards. When applicable, manufacturers can certify vehicles to either non-methane or total hydrocarbon standards.

Year	Mileage	Hydrocarbons non-methane	total	CO	NO_x
1980	50,000 miles	0.39 g/m	0.41 g/m	9.0 g/m	1.5 g/m
	100,000	0.46		10.6	1.5
1981	50,000	0.39	0.41	3.4	1.5
	100,000	0.46		4.0	1.5
1982–83	50,000	0.39	0.41	7.0	1.5
	100,000	0.46	—	8.3	1.5
1984–88	50,000	0.39	0.41	7.0	1.0
	100,000	0.46	—	8.3	1.0
1989–92	100,000	0.46	—	8.3	1.0

Diesel passenger cars and light-duty trucks are subject to the following 50,000 mile particulate exhaust standards. Diesel vehicles are subject to the particulate standards at 100,000 miles for the low-emission vehicle categories. Medium-duty vehicle particulate standards vary according to the test weight classification and low-emission vehicle category of the vehicle. For further information, please see the medium-duty vehicle section.

Year	Category	PM
1982–84		0.6 g/m
1985		0.4
1986–88		0.2
1989 and later		0.08

Fig. 8.17 Californian emission limits

- Trap: This device is used to determine the different evaporative emissions of passenger cars fitted with spark-ignition engines. The SHED method is used for this test.

Figure 8.18 shows details of the applicable emission standards [8.13] [8.17].

In Japan, fuel consumption is measured directly when the 10-mode test cycle is run, in contrast with U.S. practices that calculate fuel consumption from emission figures. The

8.3 Pollutant limits and maximum fuel consumption values

Passenger vehicles with SI or Diesel engines ≤ 10 persons

Vehicle size	Test method	Pollutant	Dim.	Limits max.	Limits mean
SI engine without weight limitation	11-mode	HC	g/T	9,50	7,00
		CO		85,00	60,00
		NO$_x$		6,00	4,40
	10-mode	HC	g/km	0,39	0,25
		CO		2,70	2,18
		NO$_x$		0,48	0,25
	Idle	HC	ppm	max.1200	
		CO	Vol.%	max.4,5	
	Falle	Evap.	g/T	2,0	
	KGH emissions		-	0	

Diesel ≤1265 kg	10-mode	HC	g/km	0,62	0,40
		CO		2,70	2,10
		NO$_x$		0,98	0,70
Diesel >1265 kg		HC		0,62	0,40
		CO		2,70	2,18
		NO$_x$		1,26	0,90
all	3-mode free accel.	Smoke blackening filter paper	%	50	

Fig. 8.18 Japanese emission standards

1. 11-mode cycle

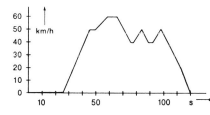

Cycle length: 1.021 km
No. of cycles/test: 4
Test length: 4.084 km
Cycle duration: 120 sec
Mean cycle speed: 30.9 km/h (39.1 km/h)*
Max. speed: 60 km/h
*without idle periods (idle content = 21.7%)

11-mode cold test: Complete 11-mode cycle 4 times; all 4 cycles are measured. 25 sec idle after cold start. Gears to be used are specified if 3- or 4-speed transmissions are used. For special transmissions, the gears to be used will be specified individually. Automatic transmissions must always be in "Drive" position. Exhaust gas analysis with CVS system.

2. 10-mode cycle

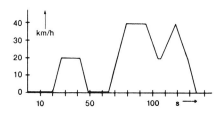

Cycle length: 0.664 km
No. of cycles/test: 6
Test length: 3.98 km
Cycle duration: 135 sec
Mean cycle speed: 17.7 km/h (24.15 km/h)*
Max. speed: 40 km/h
*without idle periods (idle content = 26.7%)

10-mode hot test: Complete 10-mode cycle 6 times; only the 5 last cycles are measured. Preconditioning: approx.15 min at 40 +/- 2 km/h, then perform exhaust gas test at idle, followed by approx. 5 min at 40 +/- 2 km/h. Gears to be used or selector position (of automatic transmissions) as for 11-mode cold test. Exhaust gas analysis with CVS system

Fig. 8.19 Japanese test cycle [8.13]

Japanese fuel consumption limits should only be regarded as guidelines, however [8.18]. Test cycles and the relevant data are found in *Fig. 8.19*.

8.3.4 European Economic Community

- **ECE regulations**

  Initial ECE regulations (ECE-R 15) were enacted in 1971 to limit CO and HC emissions. In a second stage, the admissible emissions were reduced further by 20% for CO and by 15% for HC in ECE-R 15/01 in 1975. In 1977, NO_x emissions were also limited for the first time by ECE regulation ECE-R 15/02. Further reductions of the admissible limits were introduced in 1979 (ECE-R 15/03). In the meantime, regulation ECE-R 15/04 has been enacted and is applicable to all vehicles first registered after August 1, 1985. This standard also included a number of modifications of test strategies, especially the adoption of the U.S. exhaust sampling and analysis method (CVS method) [8.16] [8.17].

- **EEC regulations**

  In 1970, regulation 70/220/EEC was introduced as the first EEC regulation to limit motor vehicle emissions. In the subsequent 20 years, this regulation was modified seven times and is now applicable to all vehicles in the form of regulation 88/436/EEC. Regulation 89/458/EEC was passed in 1989 for vehicles with a displacement of less that 1.4 liters, and it is intended to implement more restrictive limits for these vehicles from January 1, 1993. In the very first EEC regulation, a driving cycle was defined to take the special European traffic conditions into account, especially traffic in inner-city areas.

  Regulation 88/436/EEC passed in 1988 focuses on limiting emissions of air-polluting particulate emissions of diesel engines. The date of introduction for type approval of all manufactured vehicles was October, 1990. Later introduction dates, however, are applicable to direct-injection diesel engines.

  Denmark, Finland, Norway, Austria, Sweden and Switzerland (the "Stockholm group") have assumed a special position within the EEC emission regulations. They consider the EEC limits to be too lenient and have therefore adopted the more stringent U.S. standards [8.20].

  No regulations designed to limit fuel consumption have so far been passed in the European Community. The member states of the "Stockholm group", however, have adopted the fuel economy figures based on the CAFE standard along with the introduction of the U.S. standards [8.20].

8.3.5 Federal Republic of Germany

Initial exhaust emission standards were introduced in the Federal Republic of Germany in 1971. The aim of these regulations was to reduce pollutants (carbon monoxide, hydrocarbons, nitrogen oxides, lead and odorous substances) in the exhaust of spark-ignition engined motor vehicles gradually to one tenth of the average 1969 figures by 1980.

As an initial regulation, Annex XIV to §47 of the German Road Traffic Registration Law (StVZO) limited CO and HC emissions in accordance with ECE regulation R 15 from August 1, 1971. This Annex also limits smoke and soot emissions of diesel passenger vehicles in accordance with ECE regulation R 24/03. The move by the German government to introduce the U.S. standards within the EEC from January 1, 1986, however, failed due to opposition from France, Italy and Great Britain.

8.3 Pollutant limits and maximum fuel consumption values 163

Type Approval limits(T) for low-emission and
inherently low emission passenger vehicles

Exhaust level	Low-emission to Annex XXIII**	Low-emission to Annex XXV	Inherently low-emission to Annex XXIV C
Scope of application	Pass. cars* V_H all	Pass. cars*** $1.4 l \leq V_H \leq 2.0 l$	Pass. cars $V_H < 1.4 l$
Test method	FTP75 HDC	ECE 15	ECE 15
CO	2.1 g/km –	<30 g/test (<36 g/test)	38.25 g/test
HC	0.25 g/km –	–	–
NO_x	0.62 g/km –	–	6 g/test
$HC + NO_x$	– –	<8 g/test (<10 g/test)	12.75 g/test
$HDC - NO_x$	– 0.76 g/km	–	–
Particulates	0.124 g/km –	–	–
Evaporation	2.0 g/test –	–	–

( ) = Production limits

* $V_H < 1.4$ liters rated as inherently low-emission for tax legislation, conforms to Annex XXIV C

** Limits require 80,000 kms continuous operation (DL) Alternate use of impairment factors (without DL) possible

*** Also applicable to Diesel vehicles with $V_H = 1.4 l$

Fig. 8.20 Excerpt of emission standards applicable in the Federal Republic of Germany

Leading regulations passed in 1985 by the EEC Council of Ministers of the Environment enabled the government of the Federal Republic of Germany to pass more stringent emission standards. As a result, Annex XXIII to §47 of the StVZO optionally adopts the standards of the 1983 U.S. standard, and Annex XXV to §47 of the German StVZO adopts the standards of EEC amending regulation 83/351/EEC (modified in 1988 by EEC regulation 88/436/EEC) as national law. As an incentive to promote conversion of in-use vehicles, three pollutant emission classifications coupled with different tax relief levels were introduced. These classifications are specified in Annex XXIV to §47 of the German StVZO and are based on EEC regulation 83/351/EEC. *Figure 8.20* shows an excerpt of the current standards [8.13] [8.17].

Since the introduction of three-way closed-loop catalytic converters required the availability of unleaded fuel, the respective laws were passed to introduce this type of fuel as of July 1, 1985.

Neither Germany nor the member states of the EEC specify legal fuel economy requirements. Within the scope of establishing taxes on environmentally hazardous substances, plans are discussed to increase fuel tax by a surcharge for increased CO_2 emissions. As CO_2 emissions are linked directly to fuel consumption, this measure also has an effect on fuel consumption.

Annual exhaust test (ASU)

- **ASU 1**
  The German annual exhaust test (ASU) was introduced to monitor pollutant emissions of in-use motor vehicles on a regular basis. The ninth modification to the German Road

Traffic Registration Law (StVZO) stipulated an annual exhaust test that became mandatory for all vehicles with spark-ignition engine as of April, 1985. Low-emission vehicles are exempt from this test.

- **ASU 2**
  It is planned to extend the annual exhaust test to low-emission vehicles [8.21]. For this reason, an annual exhaust test for passenger vehicles with closed-loop catalytic converter and for diesel vehicles as well as for vehicles fitted with open-loop catalytic converters will therefore be introduced in 1993. On vehicles with closed-loop catalytic converter, the CO content at idle and at raised engine rpm settings, the λ control circuit and the air/fuel ratio are checked.

In other countries such as the USA, this or similar operational checks of emission control systems are already part of the vehicle certification procedure and are covered by the 50,000-mile or 5-year test of passenger vehicles with spark-ignition engines and the 100,000-mile or 10-year test of vehicles with diesel engines.

8.4 Outlook on future developments in emissions and consumption laws

8.4.1 USA – 49 states

8.4.1.1 Emission limits

- **CO emissions:** The current U.S. carbon monoxide limits for passenger vehicles are fixed at 3.4 gpm and apply both to the certification procedure and to a 50,000-mile in-use test run.

The reduction of carbon monoxide emissions is considered to be extremely important in the USA. Two CO standards linked to air quality were established by the EPA: a 1-hour limit of 35 ppm and an 8-hour limit of 9 ppm. Since at least one of these limits was exceeded in 52 U.S. regions with a total population of 87 million in the past, the primary aim of the EPA is to reduce these CO emissions that are caused primarily by motor vehicle road traffic. The following measures designed to accomplish this reduction are currently being discussed [8.22]:

- Reduction of CO limit (future standard is dependent on development of air quality),
- Introduction of a cold-temperature test procedure with corresponding limits (at low temperatures, emissions are up to 60% higher than the figures measured with previous methods). One proposal limits CO emissions at a temperature of 40 °F to 50% of the amount that vehicles first registered between 1981 and 1985 emitted at that temperature.

Another proposal plans to reduce CO emissions by 50% at a temperature of 20 °F, i.e. the vehicles must be designed to meet the current limit of 3.4 gpm at 20 °F.

- Increasing the oxygen content in the fuels.
  The current certification procedure limit of 3.4 gpm will be raised to 4.2 gpm for 100,000-mile in-use test run in 1994 [8.22].
- **HC emissions:** The current U.S. hydrocarbon emissions limit is 0.41 gpm and applies both to the certification procedure and to the 50,000 in-use test run.

This limit will eventually be modified as a matter of course as soon as a non-methane hydrocarbon limit is introduced. Methane is present in the exhaust as the hydrocarbons contain approx. 15% methane, but due to its limited reactivity, it does not contribute to

8.4 Outlook on future developments in emissions and consumption laws

ozone formation to any significant degree. For this reason, a non-methane HC limit (NMHC) will be used in the USA from 1993 to provide better assessment of the ozone-formation potential. The future U.S. limit for non-methane hydrocarbons will be 0.25 gpm. This limit will be phased in gradually from 1994 (40% of a manufacturer's sales will be required to certify to this standard from this date) and will end in 1996 when a total of 100% of a manufacturer's sales will have to certify to this standard. With this phase-in schedule, the EPA basically adopts the implementation schedules that have been planned in California for some time. This limit, however, will only be applicable to the certification procedure. A higher limit will be permitted for the durability test over the planned 100,000-mile period. This limit is currently still being discussed. Both the future Californian NMHC limit of 0,31 gpm and a value of 0.41 gpm have been considered [8.22].

- **NO_x emissions:** The current limit both for spark-ignition and for diesel engines is set at 1.0 gpm.

The nitrogen oxide emissions limit will be reduced by 60% in the USA to a future limit of 0.4 gpm. The new limit will be introduced over the same period as the HC limit [8.23].
Figure 8.21 gives an overview of the applicable current and medium-term limits.

Further drastically tightened emission limits will be implemented in the USA from Model Year 2004. This will happen if, by the end of 2001, more than 11 out of 27 cities that are subject to a particularly high ozone load still fail to meet the current ozone limits. If this should be the case, it has been announced that the HC, CO and NO_x emission limits will be lowered by 50% ("Tier 2 Standard") [8.23].

If the air pollutants in one of these cities exceed the quality standards in the year 2000 by more than 25%, all new vehicles sold in this region will also have to meet the "Tier 2 Standard" from 2004.

Before the "Tier 2 Standard" can become effective from Model Year 2004, the EPA will have to submit alternative methods and measures to the US Congress that will allow exhaust emissions to be reduced. Unless this law is amended by Congress, the "Tier 2 Standard" will become effective as soon as one of the above criteria is met [8.23].

Additional planned requirements for spark-ignition and diesel engines are highlighted in *Fig. 8.22*. The indicated values apply to different vehicle categories, these essentially being vehicles with improved emission control technology, methanol vehicles and "clean fuel" vehicles.

8.4.1.2 Testing methods

The U.S. FTP-75 test method incorporating City and Highway cycles will also remain unchanged in the near future. As described above, the durability test run was extended to 100,000 miles within the scope of the certification procedure from the date of introduction

Category of vehicles	Service life (miles)	NMHC* [g/mile]	CO [g/mile]	NO_x [g/mile]
All passenger vehicles	50.000	0,41 (0,25)	3,4	1,0 (0,4)
All commercial vehicles	100.000	0,31	3,4	–
Limits in brackets must be met as follows: MY1994: 40%, MY1995: 80%, MY1996: 100% of manufacturer's sales *Methane-free-HC limit				

Fig. 8.21 US limits related to mileage

166   8 Laws regulating the emission of pollutants and maximum fuel consumption

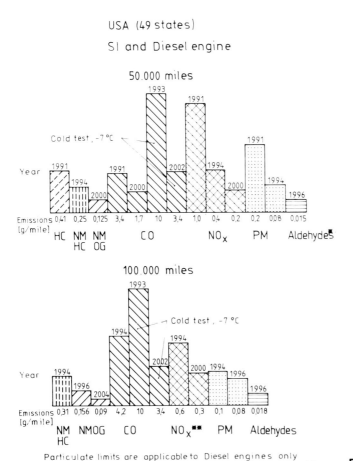

Fig. 8.22 Pollutant limits for spark-ignition and diesel engines

of the 1993 limits. For field test purposes (e.g. surveillance tests), however, the new standards will only apply up to a mileage of 75,000 miles [8.13].

8.4.1.3 Maximum fuel consumption and other regulations

Future legislation does not include plans to introduce more stringent fuel economy limits. Corporate average fuel economy figures of 30 mpg are discussed for 1995, and 45 or 55 mpg, respectively, are planned for the year 2010 [8.38]. Vehicle manufacturers may exceed the present fuel consumption limits within the scope of a "credit program" if "fuel-efficient" and "clean" vehicles (vehicles suitable for alternative fuels) are offered at the same time.

This procedure is referred to as a "banking and trading" program and applies to exhaust emissions in a comparable manner. The credit program was enacted as part of a comprehensive legislation program in early 1990. Further regulations passed by the U.S. Senate are [8.22] [8.23]:

- Vehicle manufacturers must fit all their vehicles with emission monitoring systems that provide a visual display e.g. of emission control system faults and store a fault in a suitable system (OBD II).

- All vehicle manufacturers have to fit a special container to catch fuel emissions during refueling (OBVR).
- The specific oxygen content of the fuel has to be increased (e.g. by adding ethanol or methanol) in areas that are exposed to particularly high CO concentrations.

An important step towards comprehensive emission control was made when a program to introduce "clean" fuel was passed. This program is subdivided into two phases:

- **Phase I**
The program will begin with 1995 Model Year vehicles in 9 metropolitan areas (cf. *Fig. 8.12*) that have to cope with particularly high ozone concentrations. In these cities, the ozone-producing emissions of any newly-sold vehicle must not exceed a limit of 0.75 gpm. Emissions of today's vehicles, however, currently average 2.8 gpm. The initial phase also includes regulations to reduce toxic emissions by 12%.

- **Phase II**
This will begin in Model Year 1999 and introduces more stringent standards to replace the standards introduced in Phase I. Ozone-producing emissions must not exceed a limit of 0.66 gpm in the field. In addition, emissions of air-polluting substances must be reduced by 27% over the emissions of conventionally operated vehicles. The EPA reserves the right to replace the 27% reduction by a 18% reduction if it is found that the more stringent limits cannot be attained with available engineering. This stage also assumes that a reduction of emissions can be accomplished virtually exclusively by using alternative fuels.

The U.S. States have been granted certain allowances for further action. These are:

- The introduction of alternative fuels may be completed regardless of nation-wide schedules.
- The individual States may increase the minimum percentage of "clean" vehicles sold if required.
- The States may extend the program to further urban areas in addition to the 9 specified cities.
- The State of California is granted the right to enact a separate program for introduction of alternative fuels. Other States may adopt this regulation.
- Manufacturers of mineral oils are to be encouraged to introduce alternative fuels for distribution from existing filling stations.

8.4.2 USA – California

8.4.2.1 Emission limits

Comprehensive modifications of emission limits have been passed and/or planned by the California Air Resources Board (CARB) for the future. These measures are designed to meet the demands for reduction of ozone-generating emissions as set forth in the Air Quality Management Plan (AQMP). The measures proposed by the CARB cover two phases, i.e. a "long-term" and a "short-term" program. The proposals described in the "short-term" program have to be introduced by 1993 while the framework described in the "long-term" program is to be introduced over the 1994 to 1997 period [8.24]. No new limits have been established within the "short-term" program, and our description will therefore focus on the "long-term" program.

One important modification concerning definition of pollutants is the abovementioned specification of non-methane hydrocarbon limits referred to as OMHCE (Organic Material Hydrocarbon Equivalent) for methanol-operated vehicles and as NMOG (Non-Methane Organic Gas) for "clean-fuel" vehicles.

During an initial two-year period, additional "intermediate standards" that are relatively easy to comply with will be applicable to tests of in-use vehicles ("In-Use Compliance Test") in order to make it easier for motor vehicle manufacturers to prepare for the more stringent standards. These "intermediate standards" for passenger vehicles are applicable over a period of 50,000 miles with a limit of 0.32 gpm for NMHC and 5.2 gpm for CO [8.25].

Various models have been proposed for phasing in all the above standards. According to the implementation models shown in *Fig. 8.23*, the conversion of all vehicles to the new, more stringent standards would be completed by 1997 or 1995, respectively.

The percentages of both figures refer to total vehicle sales. A formaldehyde emission limit of 15 mg per mile will apply to methanol vehicles (including flexible-fuel vehicles) that will be certified as of 1993 [8.25].

The "long-term" program of the Californian Air Resources Board also plans to introduce very low-emission vehicles that will comply with correspondingly more stringent standards as of Model Year 1994. This above-average emission control level is to be achieved primarily by using "clean" fuels (e.g. methanol, ethanol). The use of energy sources that are no longer based on hydrocarbons (e.g. electrical energy, hydrogen) is also planned as these fuels will allow "zero emissions" of HC, CO, CO_2, NO_x etc. to be achieved. According to the pollutant quantities emitted and to the introduction time frame, these vehicles are classified as follows [8.13]:

- **TLEV:** Introduction of the TLEV category (Transitional Low-Emission Vehicles) starts in 1994 and has the primary aim of reducing emissions of ozone-generating substances. Some of the vehicles certified today are already capable of meeting the TLEV standards without any substantial engineering modifications if they are operated with "clean" fuels.
- **LEV:** TLEV will be replaced by "Low-Emission Vehicles" (EV) in 1997. These standards correspond to the considerable advance of engineering to be expected and can be met by introducing new emission control technologies (e.g. preheating of the catalytic converter), reduction of raw emissions and use of alternative fuels.
- **ULEV:** Introduction of the ULEV (Ultra-Low Emission Vehicles) will start in 1997, but will only become mandatory for higher shares of overall sales from the year 2000. The

Introduction Model 1

Model Year	Comply with Current Standards	Certify to Proposed Standards	Comply with Intermediate In-Use Stds	Comply with Proposed In-Use Stds
1993	60%	40%	40%	-
1994	20%	80%	80%	-
1995	-	100%	60%	40%
1996	-	100%	20%	80%
1997	-	100%	-	100%

Introduction Model 2

Model Year	Comply with Current Standards	Certify to Proposed Standards	Comply with Intermediate In-Use Stds	Comply with Proposed In-Use Stds
1993	60%	40%	40%	-
1994	20%	80%	80%	-
1995	-	100%	-	100%

Fig. 8.23 Models for implementation schedule of different standards [8.25]

8.4 Outlook on future developments in emissions and consumption laws

1993 and later Gasoline, Diesel and Methanol passenger cars

Manufacturers must certify a minimum of 40% of their 1993, 80% of their 1994 and 100% of their 1995 and later passenger cars plus light-duty trucks to the primary or low-emission vehicle standards, with the remainder certifying to the secondary standards. 1993 vehicles certified to the 0.7 g/m NO_x standard are subject to a 7 year/75,000 mile recall for selected emission control parts. Manufacturers choosing to certify diesel passenger cars to the optional standards must demonstrate compliance at 100,000 miles. For methanol-fueled vehicles, including flexible-fueled vehicles, NMHC means organic material hydrocarbon equivalent (OMHCE). Beginning in model year 1994, manufacturers are also required to meet a fleet average non-methane organic gas (NMOG) requirement. See additional standards listings for other emission requirements.

Year	Mileage	NMHC	CO	NO_x	Notes
1993–94 primary	50,000 miles	0.25 g/m	3.4 g/m	0.4 g/m	
	50,000	0.25	3.4	0.7	
	100,000	0.31	4.2	—	1993 option only
secondary	50,000	0.39	7.0	0.4	
	50,000	0.39	7.0	0.7	optional
	100,000	0.46	8.3	1.0	diesel option
1995 and later primary	50,000	0.25	3.4	0.4	
	100,000	0.31	4.2	—	
	100,000	0.31	4.2	1.0	diesel option

Low-Emission Vehicle Categories

Low-emission vehicles are vehicles operating on any fuel that have been certified by the Air Resources Board to meet the following exhaust emission standards. These emission standards are used to compute the fleet average NMOG. The emissions of alternate fueled vehicles may be adjusted to account for the lower reactivity of the NMOG emissions. Flexible-fuel and dual-fuel vehicles must also certify to the gasoline standards.

50,000 mile standards	NMOG	CO	NO_x
Low-Emission Vehicles			
Transitional Low-Emission Vehicle (TLEV)	0.125 g/m	3.4 g/m	0.4 g/m
Low-Emission Vehicle (LEV)	0.075	3.4	0.2
Ultra Low-Emission Vehicle (ULEV)	0.040	1.7	0.2
Zero Emission Vehicle (ZEV)	zero	zero	zero
Gasoline standards for flexible and dual fuel Low-Emission Vehicles			
Transitional Low-Emission Vehicle (TLEV)	0.25 g/m	3.4 g/m	0.4 g/m
Low-Emission Vehicle (LEV)	0.125	3.4	0.2
Ultra Low-Emission Vehicle (ULEV)	0.075	1.7	0.2

100,000 Mile Standards	NMOG	CO	NO_x
Low-Emission Vehicle standards			
Transitional Low-Emission Vehicle (TLEV)	0.156 g/m	4.2 g/m	0.6 g/m
Low-Emission Vehicle (LEV)	0.090	4.2	0.3
Ultra Low-Emission Vehicle (ULEV)	0.055	2.1	0.3
Zero Emission Vehicle (ZEV)	zero	zero	zero
Gasoline standards for flexible and dual fueled Low-Emission Vehicles			
Transitional Low-Emission Vehicle (TLEV)	0.31 g/m	4.2 g/m	0.6 g/m
Low-Emission Vehicle (LEV)	0.156	4.2	0.3
Ultra Low-Emission Vehicle (ULEV)	0.090	2.1	0.3

Fleet average NMOG requirements

In addition to the individual model year exhaust emission standards, a manufacturer's fleet average NMOG emissions must not exceed the following levels. The fleet average requirements apply to the manufacturer's combined fleet of passenger cars and light-duty trucks (0-3750 lbs.). Compliance with the fleet average requirements is meet by averaging the NMHC or NMOG standards of vehicles certified to the primary, secondary or optional standards with vehicles certified to the low-emission vehicle categories. In order to receive credit for the lower NMOG of a low-emission vehicle category, a vehicle must meet the CO and NO_x standard for the category to which it is certifying. NMOG emissions include oxygenated and non-oxygenated hydrocarbons. Beginning in 1998, a minimum percentage of each manufacturer's combined sales of passenger cars and light-duty trucks (0-3750 lbs.) will be required to be zero-emission vehicles.

Year	NMOG	Percentage of ZEV's required
1994	0.250 g/m	
1995	0.231	
1996	0.225	
1997	0.202	
1998	0.157	2%
1999	0.113	2%
2000	0.073	2%
2001	0.070	5%
2002	0.068	5%
2003 and later	0.062	10%

Fig. 8.24 Overview of future limits

1993 and later Formaldehyde and Diesel particulate Matter (PM) standards	Manufacturers of methanol and flexible fueled passenger cars, light-duty trucks and medium-duty vehicles must comply with the following formaldehyde standards at 50,000 miles. The standards are in milligrams per mile.

Vehicle Type	Weight (lbs.)	Formaldehyde
Passenger cars	All	15 mg/m
Light-duty truck and	0–3750	15

Low-Emission Vehicle Formaldehyde Exhaust Emission Standards	To be certified by the Air Resources Board as a low-emission vehicle, passenger cars, light-duty trucks and medium-duty vehicles must also meet the following formaldehyde standards. The standards are in milligrams per mile.

Vehicle Type	Weight (lbs.)	Mileage (miles)	Category	Formaldehyde
PC and LDT	all 0–3750	50,000	TLEV	15 mg/m
			LEV	15
			ULEV	8
		100,000	TLEV	18
			LEV	18
			ULEV	11

Diesel passenger cars and light-duty trucks are subject to the following 50,000 mile particulate exhaust standards. Diesel vehicles are subject to the particulate standards at 100,000 miles for the low-emission vehicle categories. Medium-duty vehicle particulate standards vary according to the test weight classification and low-emission vehicle category of the vehicle. For further informaton, please see the medium-duty vehicle section.

Year	Category	PM
1982–84		0.6 g/m
1985		0.4
1986–1988		0.2
1989 and later		0.08
1993 and later	TLEV	0.08
	LEV	0.08
	ULEV	0.04

Fig. 8.25 Overview of future limits

TLEV limits can only be complied with by introducing new emission control technology and using alternative fuels as well as by reducing raw emissions yet further since this category stipulates both a reduction of CO emissions by another 50% and a further reduction of NMHC emissions.
- **ZEV:** A further intention of the CARB is the ZEV (Zero-Emission Vehicles) category. This means that from 1998, an emissions standard of NMOG = 0.0 gpm (NMOG = Non-Methane Organic Gas) must be complied with by 2% of the vehicle fleet. This limit must be complied with by 5% in 2001 and by 10% of the fleet in 2003.

It is also planned to phase in TLEV, LEV and ULEV vehicles gradually. With the exception of ZEV figures, each manufacturer may select the rates individually as long as the standard corporate mean value is complied with. Starting with a NMOG value of 0.25 gpm in 1994, this value will be reduced to 0.062 gpm in 2003.

An overview of future limits that result from the combination of all measures including TLEV, LEV, ULEV and ZEV is shown in *Fig. 8.24* and *Fig. 8.25*.

8.4.2.2 Testing methods

No changes to the current U.S. FTP-75 test procedure have been planned for California. Within the scope of the certification procedure, however, the durability test will be extended from 50,000 to 100,000 miles once new HC and CO standards are introduced for all vehicles in 1993. For real-life tests (e.g. "In-Use Compliance Test" or "Surveillance Test"), however, the new standards will only be applicable to a maximum mileage of 75,000 miles. Furthermore, the current U.S. test procedure does not include tests according to the Californian "Inspection and Maintenance" program (Smog Check). In the past, some occurrences were reported where specific vehicles failed the smog checks although they did meet U.S. certification standards. From Model Year 1993, passing of this test will therefore become a compulsory part of the Californian certification test for passenger vehicles [8.11].

The additional cold start test planned by the EPA as a means of reducing carbon monoxide emissions by a significant degree (refer to Section 8.4.1.1) is currently not yet being discussed in California. If, however, the EPA should decide to incorporate a separate test of this type into the certification procedure, it may be assumed that this test will also be adopted by the Californian CARB authority.

8.4.2.3 Maximum fuel consumption and other regulations

No moves to introduce more stringent fuel consumption standards beyond the current standards are currently being planned in California [8.26]. While the above "short-term" program does not aim to tighten down the permissible pollutant emission limits, the fuel oil industry will probably have to comply with the following regulations on fuel quality [8.24]:

- Reduction of benzene content in gasoline from 2.0% by volume to 0.8% by volume.
- Reduction of volatility rate by modifying the Reid vapor pressure (RVP) from 10.0 psi to 8.0 psi.
- Introduction of additives to prevent deposit formation in engine combustion chambers.
- Increasing the oxygen content of fuels during the winter months,
- Reduction of content of aromatic substances in the fuel.

The "long-term" program plans to introduce further measures to reduce pollutant emissions of passenger vehicles. For one thing, the fuel oil industry will be forced to increase the production and sale of alternative fuels in accordance with the planned introduction dates of TLEV, LEV, ULEV and ZEV vehicles [8.24].

Furthermore, the automotive industry will have to meet additional requirements pertaining to certification of new vehicles from Model Year 1994. As part of these requirements, it is planned to introduce an "On-Board II" diagnostic system that performs a wide range of monitoring tasks. By the 1996 Model Year, the below monitoring features will have to be implemented in all motor vehicles [8.11]:

- Monitoring of catalytic converter efficiency. A malfunction message has to be displayed if the conversion efficiency drops below the 50% threshold,
- Display of ignition failure or misfire.
- Display of malfunctions of the evaporative emissions recirculation circuit from the activated carbon system to the engine.
- Monitoring of operation of secondary air injection.
- Monitoring the CFC circuit of the air conditioning system for leaks. Manufacturers may omit this monitoring system if they use only non-CFC air conditioning systems by the 1996 Model Year.

- Display of fuel delivery irregularities.
- Display of operability of the oxygen (lambda) sensor.
- Monitoring of the exhaust recirculation system. The recirculation rate must be measured continuously so that a warning lamp will light up if the actual value exceeds or falls below a specified standard.
- Functional check of all emission-related components.
- Protective measures covering all monitoring functions in order to avoid any tampering with the on-board computer.

The malfunctions displayed do not necessarily mean that the respective components are completely inoperative but are rather intended to indicate to the driver that optimum emission control is no longer ensured.

If a malfunction is displayed, the driver must have his/her car checked by a repair shop in order to have the malfunction repaired at the expense of the manufacturer. Since all informations are at the same time stored in a central computer, a check in the shop will immediately reveal any other irregularities that may not yet have been displayed. If this monitoring system proves its worth, the Californian "Inspection and Maintenance Program" may even be dispensed with in the long run [8.11].

The Californian Environmental Agency also plans to introduce the "credit program" for motor vehicle manufacturers as a means of accelerating the introduction of very low-emission vehicles (LEV, ULEV, ZEV). If a manufacturer sells more low-emission vehicles from 1993 than he is required by law (this covers LEV and ULEV vehicles only), this manufacturer is granted bonus credits according to a highly complex awarding system. These credits will allow the manufacturer to market other vehicles within his range that would exceed the specified emission limits. The manufacturers will also be allowed to trade such bonus credits among each other [8.25].

The same system also applies to fuel suppliers. They will also be given credits if they produce a larger quantity of alternative fuels than would be demanded by law.

8.4.3 Japan

8.4.3.1 Emission limits

Extensive recommendations for implementing more stringent emission standards were published by the Japan Environmental Agency in December, 1989. It is planned that these recommendations will be adopted by the responsible "Ministry of Transport". In brief, the following modifications are recommended for passenger vehicles, especially for diesel-engined passenger vehicles [8.27]:

- Reduction of emission standards for nitrogen oxides of diesel-engine passenger vehicles in two stages to 0.4 grams per km (gpk).
- Introduction of a particulate emission limit for diesel-engine passenger vehicles from 1994. The planned limit of 0.2 gpk will be lowered by another 60% in a second stage. Particulate emissions will be determined according to the EPA analysis process.
- Reduction of emitted smoke quantity in two stages by 50%.

Figure 8.26 shows the standards for future limits.

8.4.3.2 Testing methods

In the recommendations submitted in December, 1989, the Japan Environmental Agency has recommended modifications to the existing 10-mode driving cycle. One main area of

8.4 Outlook on future developments in emissions and consumption laws

Category of vehicles	Kind of emissions	Existing standards		New Measurement mode	Short-term target (Phase 1)			Long-term target (Phase 2)	
		Mean values	Year of enforcement		Target mean value	Scheduled year of enforcement	Reduction rate	Target mean values	Reduction rate
Diesel powered Passenger vehicles	NO$_x$	EIW ≤1,25t 0,7 g/km max. 0,98	1986 (manual transmission vehicles) 1987 (automatic transmission vehicles)	New 10-mode (10·15 mode)	0,5 g/km max. 0,72	(1994)	(29%)	0,4 g/km	(43%)
		EIW >1,25t 0,9 g/km max. 1,26	1986 (manual transmission vehicles) 1987 (automatic transmission vehicles)		0,6 g/km max. 0,84	(1994)	(33%)		(56%)
	PM	–			0,2 g/km	1994	–	0,08 g/km	60%
	Smoke (3-mode)	50%	1972	(Current 3-mode)	40%	Concurrently with PM	20%	25%	50%

Fig. 8.26 Future emission standards for Japan [8.27]

10 . 15-mode cycle

Application: Nov. 1, 1991 for new models,
and April 1, 1993 for imported vehicles

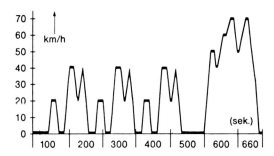

No. of cycles/test: 1
Cycle duration: 660 sec
Cycle length: 4.16 km
Mean cycle speed: 22.7 km/h
Max. speed: 70 km/h

10 . 15-mode hot test: For preconditioning, drive at 60 km/h for 15 mins. Then perform idle exhaust test, followed by 5 mins at 60 km/h and one 15-mode cycle. After preconditioning, drive combined 10 . 15-mode cycle (three 10-mode cycles and one 15-mode cycle) once and perform measurements.

Fig. 8.27 New Japan test cycle [8.13]

criticism was the maximum test speed of 40 km/h (25 mph) considered too low by the Agency. The "Japan Environment Agency" hopes that raising this speed to a maximum of 70 km/h (45 mph) will allow national driving habits to be reproduced more accurately. A new 10-mode test applicable to both spark-ignition and diesel engine passenger vehicles was introduced in 1991 [8.27]. This new test cycle is shown in *Fig. 8.27*.

8.4.3.3 Maximum fuel consumption and other regulations

No legal standards to limit fuel consumption of passenger vehicles will be introduced in Japan in the future. Medium and long-term changes of the vehicle taxing system have been discussed, however. According to this, the present tax system may be complemented or even replaced by a new fuel consumption tax.

The Japanese government hopes that the introduction of such a taxing system and of other legal regulations will reduce further demand for motor vehicles, especially for passenger vehicles [8.27] [8.28].

The use of vehicles operated with alternative fuels, however, will be limited to research and testing purposes for the time being. Concepts similar to those submitted in the USA do not exist in Japan.

In December, 1989, however, recommendations were submitted to reduce the admissible sulphur content in diesel fuel in order to reduce vehicular pollutant emissions. According to this program, the current sulphur content is to be reduced from 0.4 to 0.5% by volume to 0.05% by volume in two steps.

Further measures to be adopted in the future will concentrate on increased use of vehicles with particularly low emissions, introduction of efficient conversion concepts for in-use vehicles and on implementing special traffic-related measures (e.g. traffic routing systems) [8.27] [8.28].

8.4.4 European Economic Community

8.4.4.1 Emission limits

The development of future European emission control legislation covers two stages. The first stage is defined by regulations 88/76/EEC and 89/458/EEC passed by the Council of Europe. These regulations specify lower limits based on the current testing method (in

Vehicles with spark-ignition or Diesel engine

Displacement $V_H < 1.4 l$	CO g/test	$\Sigma HC\text{-}NO_x$ g/test	$NO_{x\,max}$ g/test	PM Particulates g/test
For Type Approval from Oct 1, 1990	45	15	6.0	1.1
For production vehicles from Oct. 1, 1991	54	19	7.5	1.4
For Type Approval from July 1, 1992	19	5.0	–	–
For production vehicles from Dec 1, 1992	22	5.8	–	–

Displacement $1.4 l < V_H < 2.0 l$	CO g/test	$\Sigma HC\text{-}NO_x$ g/test	PM Particulates g/test
For Type Approval from Oct. 1, 1991	30	8	1.1
For production vehicles from Oct 1, 1993	36	10	1.4

Displacement $V_H > 2.0 l$	CO g/test	$\Sigma HC\text{-}NO_x$ g/test	$NO_{x\,max}$ g/test	PM Particulates g/test
For Type Approval from Oct. 1, 1988	25	6.5	3.5	1.1
For production vehicles from Oct. 1, 1989	30	8.1	4.4	1.4

Fig. 8.28 Displacement-related emission standards

accordance with regulation 70/220/EEC) and subdivide the vehicles into specific categories of engine displacement. For this reason, a variety of different limits exists. The following listing shows an overview of mandatory emission standards and the respective introduction dates (*Fig. 8.28*) [8.29] [8.30] [8.31].

The emission standards that have already been introduced are valid

- both for vehicles (with spark-ignition and diesel engines) with a displacement below 1,400 c.c., and
- for vehicles with a displacement of more than 2,000 c.c.
- Particulate emissions of passenger vehicles with diesel engines were also regulated as of 1989. The type approval limit of 1.4 grams per test, however, will only be applicable to direct-injection diesel vehicles from August 1, 1994, or August 1, 1996, respectively.

In the second stage of future European emissions legislation, an extended testing method that was passed by the Environmental Council of the EEC in 1989 is to be introduced [8.32] [8.2].

A new regulation was submitted to the responsible authorities in February, 1990. In addition to the abovementioned test cycle and the respective limits, this proposal for the first time also specifies limits for evaporative emissions and the introduction of a long-term durability test.

The proposal for modification of regulation 70/220/EEC includes the below individual emission limits (*Fig. 8.29*) [8.2].

This overview shows that the proposal submitted by the Commission is based on equal standards for all displacement sizes and engine types. Only the particulate and evaporative emission standards are specific to diesel and spark-ignition engines, respectively. It should also be noted that these standards were implemented on the basis of the limits applicable to vehicles with a displacement below 1,400 c.c. in accordance with EEC regulation 89/458/EEC. This procedure was adopted to observe the decision of the EEC Environmental Council to introduce more stringent emission standards. According to statements by the Council of the European Community, these standards may be considered "at least as restrictive as those of the United States of America".

The proposal submitted by the Commission also includes an important regulation for obtaining an EEC Type Approval. Up to the present, a vehicle that was certified in accordance with US test procedure FTP-75 cannot be refused an EEC Type Approval (Annex III A). Since the opinion of the Commission is that the proposed more stringent standards combined with the new, comprehensive test procedures reflect European driving conditions better than the U.S. standards [8.2], it is intended that this Annex will be deleted after a certain transitional period.

According to the proposal submitted by the Commission, Annex III A will remain in effect for granting Type Approvals for new vehicle types for another two years, i.e. up to June 30, 1994, and for registering new vehicles up to December 31, 1995. During this

	CO g/km		$\Sigma HC + NO_x$ g/km		Partic. g/km		Evap. g/test	
T = Type approval S = Product. standard	T	S	T	S	T	S	T	S
all SI engine passenger vehicles	2,72	3,2	0,97	1,1	-	-	2,0	-
all Diesel engine passenger vehicles	2,72	3,2	0,97	1,1	0,19	0,24	-	-

Fig. 8.29 Proposal for emission standards in the EEC

Year		CO [g/km]	HC+NO$_x$ [g/km]	Partic. [g/km]
1992/93	Type Approval	2,72	0,97	0,14
1992/93	Production testing	3,16	1,13	0,18
1996	–	2,72	0,48 Otto / 0,90 Diesel	0,06 Diesel

Fig. 8.30 European emission standards

transitional period, vehicle manufacturers are free to choose between either FTP-75 of Annex III A or the new European testing procedure.

In late 1990, the Ministers of the Environment of the EEC member states agreed to introduce more stringent emission standards. As a complement to *Fig. 8.29*, the standards described in *Fig. 8.30* will be required for new models from mid-1992 and for all new vehicles from 1993. Standards to be introduced from 1996 are being discussed in a second stage [8.33].

8.4.4.2 Testing methods

As discussed in the previous chapter, a decision was made by the EEC Environmental Council on June 8, 1989, to introduce an extended test cycle. The overall driving cycle for the future European emission testing method is composed of two basic driving cycles (see *Fig. 8.2*). The city driving cycle that is designed to reproduce the driving conditions prevalent in urban areas with a high traffic density must be completed four times (as in the current test cycle specified by regulation 70/220/EEC).

The newly-defined extra-urban driving cycle is intended to represent the driving conditions prevalent on roads outside metropolitan areas and on highways. The theoretical driving pattern of the extra-urban cycle corresponds to that of the urban cycle, and its top speed is limited to 120 km/h (80 mph). The additional test cycle has a test length of 400 s and must be completed once [8.34].

Vehicles with "low-power" engines, i.e. with a power-to-weight ratio of 40 kW per 1,000 kgs or less, the maximum speed or the extra-urban test is reduced to 90 km/h (55 mph).

The complete driving cycle of the new European test procedure, however, should be considered as one single unit that must be completed without interruptions. The pollutant quantities measured should therefore be compared with limits that refer to the complete test.

In addition to a modification of the test cycle, the proposal submitted by the Commission also includes plans for introduction of a test procedure to check evaporative emissions. For this purpose, it is recommended that the current U.S. testing method (the SHED test) be incorporated into the new regulation.

The intention, however, is to use this procedure only for Type Approval testing since it is too time-consuming for production testing and would require highly complex testing equipment in the test laboratories. For production checking purposes, it will therefore be admissible to limit checks to component inspections.

The proposal submitted by the Commission also contains the requirement to monitor the service life of vehicle components and systems used to reduce emissions of gasoline vehicles (e.g. catalytic converters and exhaust recirculation systems) within the scope of the certification procedure.

The below three durability tests have been suggested for inclusion in future European emission standards [8.2]:

- The current U.S. durability test for those European motor vehicle manufacturers that export vehicles into the U.S.A.,
- A European durability test limited to 30,000 km (20,000 miles) but run under more severe conditions in order to ensure that it is on a level with the U.S. test. This test addresses manufacturers of vehicles primarily intended for the EEC market,
- A number of deterioration factors to be adopted by those automotive manufacturers who do not wish to carry out durability tests.

Both the deterioration factors that result from the durability tests and those factors that are specified directly by the regulation are, in accordance with the proposal submitted by the Commission, valid both for granting Type Approvals for new vehicle models and for conformity checks of production vehicles.

8.4.4.3 Maximum fuel consumption and other limits

The proposals submitted by the Commission of the European Community for future emission regulation do not include any steps to enforce direct limits on fuel consumption. Both regulation 89/458/EEC and the new proposed regulation, however, include an identical section that proposes measures to reduce carbon dioxide emissions of motor vehicles [8.30].

The government of the Federal Republic of Germany has asked the EEC to introduce regulations for reducing fuel consumption. This regulation should take differences in vehicle weight into account and should contribute to a reduction of CO_2 emissions. The proposals envisage the following measures:

- CO_2 emissions are measured in a newly introduced driving cycle.
- The limits are dependent on vehicle weight. The slope of the regression lines calculated from CO_2 measurement data, vehicle weights and the corresponding centrifugal masses may be modified in accordance with the year of first registration of the vehicle in order to achieve more stringent limits by 2005.

Figure 8.31 shows the CO_2 emissions as a function of curb weight and includes measurement results obtained on 128 vehicles for the 1993 limits and the 2005 targets [8.35]. It is planned to subdivide the vehicles according to their weight into weight classes up to 1500 kg, 1100 kg and 800 kg. According to this classification, the proposed limits may also be indicated as a function of fuel consumption (*Fig. 8.32*).

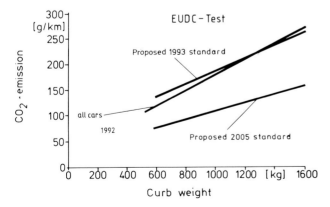

Fig. 8.31 CO_2 emissions as a function of the vehicle mass

Curb weight [kg]	CO_2-emission [g/km]		Otto engine [l/100 km]		Diesel engine [l/100 km]	
	1993	2005	1993	2005	1993	2005
800	160	96	6,9	4,1	6,1	3,7
1100	199	119	8,6	5,1	7,6	4,5
1500	250	150	10,8	6,5	9,5	5,7

Fig. 8.32 CO_2 emission limits as a function of vehicle curb weight [8.35]

The European Parliament, on the other hand, has endorsed limits established according to the engine displacements. For engines with a displacement of 2 liters, a value of 250 gpk for CO_2 has been planned, with this value dropping to 200 gpk CO_2 for displacements from 1.4 to 2 liters and to 160 gpk CO_2 for displacements below 1.4 liters.

Another regulation proposed by the Commission intends to reduce evaporative emissions generated during the transport and storage of gasoline fuels from the refinery to the filling stations. These measures are referred to as "Stage 1" and affect 5 to 7% of the anthropogenic emissions of volatile organic compounds (VOC).

8.4.5 Federal Republic of Germany

8.4.5.1 Emission limits

The regulations on limiting motor vehicle exhaust emissions passed by the EEC will also be valid in the Federal Republic of Germany as Germany is part of the EEC. The EEC regulations will therefore constitute the basis for future national exhaust emission regulations. For this reason, the EEC plans for future development of exhaust emissions legislation will also be applicable in Germany.

8.4.5.2 Testing methods

Testing methods are an essential part of the entire emission control legislation and have been stipulated in the relevant EEC regulations. The statements made in the previous section are therefore also true in this context. This means that the Federal Republic of Germany is under the obligation of adopting the test methods specified by the European Community.

8.4.5.3 Maximum fuel consumption and other limits

Limiting motor vehicle fuel consumption is a major issue, especially with regard to the rapid increase of carbon dioxide emissions and their abovementioned effect on global warming.

For this reason, both the European Community and the German Federal Ministry of the Environment plan to introduce a number of measures to reduce CO_2 emissions. A number of basic procedures are available to put these measures into practice:

- Introduction of direct fuel consumption limits according to U.S. practice. This, however, can only be accomplished by implementing relevant EEC regulations on a European scale. No plans for any such action, however, are currently known, and it is not likely that this idea will be adopted on a medium or long-term basis.

8.4 Outlook on future developments in emissions and consumption laws

- Increasing the fuel price by raising fuel tax or introducing an additional fuel tax. These measures are designed to reduce annual mileage and to encourage development of low-consumption vehicles. An advantage of this approach is that such measures can be introduced in Germany independent of the existence of corresponding EEC regulations. The amount of tax increase demanded varies widely, however [8.36].
- A third approach is based on modifying the present motor vehicle tax scheme. This possibility may also be invoked independent of any corresponding development in the European Community.

The German motor vehicle inspection organization (TÜV) has submitted proposals for an environmentally oriented motor vehicle tax system. This proposal is based on an evaluation coefficient that is designed to differentiate among the emission control concepts available on the market today. The evaluation coefficient is used to compare the sum of actual emissions to the emissions of vehicles fitted with three-way closed-circuit catalytic converters.

In this way, the "tax formula" can be based on a linear correlation that is determined by a minimum tax for low-emission vehicles and by the desired tax rate for the technology given the poorest ratings.

As an incentive to developing low-consumption vehicles and, hence, as a means to take CO_2 emissions into account, this "tax formula" may also be linked to an evaluation coefficient for fuel consumption.

From an environmental point of view, the new tax formula should create an incentive that will not only encourage compliance with a specified standard (today's limits) but will also advance the technological state of the art in order to ease the financial burden imposed by the new "ecological tax". The lower the pollutant emissions of a vehicle, the lower is the tax to be paid [8.37].

In order to determine the evaluation coefficient, the emissions of the individual vehicle models with different exhaust emission concepts will be compared to mean emissions of a vehicle with a three-way closed-loop catalytic converter. To determine the emissions, the new European driving cycle is used as this cycle will also allow typical urban, extra-urban and highway driving conditions to be taken into account. The emissions of the reference vehicles will be determined in a European driving cycle specified by law. The pollutants subject to legal restrictions, i.e. carbon monoxide, unburned hydrocarbons, nitrogen oxides and particulates, will be taken into account in order to determine the evaluation coefficient. In the case of particulate emissions, the planned Californian standards will be used as a reference unit. *Figure 8.33* shows the scatterband of the evaluation coefficients of the different pollutant reduction concepts as well as the respective mean value.

The selection of a reference vehicle – a medium-size vehicle fitted with a three-way closed-loop catalytic converter – takes the state of the art into account and implicitly accounts for differences in the toxicological assessment of the individual pollutants since the limits also reflect the probable health hazard potential of the exhaust gas components.

In order to encourage development of low-consumption vehicles and to take CO_2 emissions into account that are directly dependent on fuel consumption, the tax formula is multiplied by the relative fuel consumption. The fuel consumption of modern economical medium-size vehicles has been proposed as a reference value by the German motor vehicle inspection organization (TÜV).

Figure 8.34 shows the "tax curve" based on the evaluation coefficient.

Other "tax curves" that may in certain cases be given a progressive shape, have also been considered in order to reflect potential tax incentives designed to encourage acquisition of a vehicle fitted with emission control systems. Any decision on the progression of this straight line will eventually be dictated by fiscal considerations since the eventual tax

180   8 Laws regulating the emission of pollutants and maximum fuel consumption

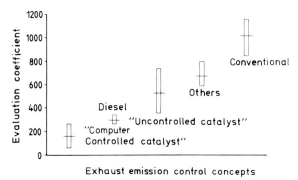

Fig. 8.33 Evaluation coefficients of individual pollutant reduction concepts [8.37]

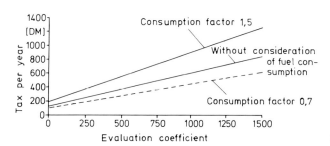

Fig. 8.34 Tax curves [8.37]

Country	Year	HC	NOx	CO	PM	Durability
USA/49 States FTP 75		gpm				
	1992 (Current)	0,41	1,0	3,4	0,2	50 kmiles
	1994/95/96 40/80/100%	0,25/0,31 (NMHC)	1,0/1,25	3,4/4,2	0,08/0,10	50kmiles / 100kmiles
	2001	0,125 (NMHC)				50 kmiles
California FTP 75		gpm				
	1992 (Current)	0,39 (NMHC)	1,0	7,0	0,08	100 kmiles 50 kmiles for PM
	1994/95/96 40/80/100%	0,31 (NMHC)	1,0	4,2	0,08	100 kmiles 50 kmiles for PM
	TLEV	0,125/0,156	0,4/0,60	3,4/4,2	-/0,08	
	TEV	0,075/0,090	0,2/0,30	3,4/4,2	-/0,08	50/100 kmiles
	ULEV	0,040/0,055 (NMOG)	0,2/0,30	1,7/2,1	-/0,04	PM 100kmiles
Europe ECE 15 + EUDC		g/km HC + NOx				
	1992 (Current) (I)	0,97		2,72	0,14	80·10³ km
	MVEG Proposal 96/IDI (II) 99/IDI	0,7/IDI 0,9/DI		1,0/IDI 1,0/DI	0,08/IDI 0,10/DI	80/160·10³ km?
	Proposal Germany 2000 (III)	0,5		0,5	0,04	80/160·10³ km?
Japan Japan 10.15 cycle *inertiaweight(test)		g/km				
	1992 (Current)	0,40	0,5/<1250kg* 0,9/>1250 kg	2,10		
	93/94	0,40	0,5/<1250kg 0,6/>1250kg	2,10	0,20	
	2000	0,40	0,40	2,10	0,08	

Fig. 8.35 Emission control regulations [Source: AVL]

revenue will obviously decrease as the number of "clean" passenger vehicles increases [8.37].

This tax formula also takes fuel consumption into account and shows the respective progression lines of a vehicle consuming 50% more than the reference vehicle (consumption factor 1.5) and of a vehicle consuming 25% less (consumption factor 0.75).

Pollutant emissions and fuel consumption required to establish the total tax due are determined in the Type Approval test on the basis of the new European driving cycle.

When the proposed tax system is linked to emissions and fuel consumption figures determined during the Type Approval test, production emission testing of new vehicles and the annual exhaust test mandatory in Germany would become increasingly important.

Again, however, it will not be possible to introduce a system of this nature on a short-term basis since the new tax system will have to be discussed and harmonized on a political level, both nationally and internationally.

Figure 8.35, finally, provides an overview of the most important emission control regulations and proposals for the regions considered above, i.e. USA (49 States), California, Europe and Japan (current as of 1993).

References

Chapter 1

[1.1] Ahrens, G.A.; Becker, K.: Umwelt- und Verkehrs-Maßnahmen zur Minderung der Umweltbelastungen des Verkehrs, VDI Fortschrittberichte Nr. 150, S. 109, 1991
[1.2] Metz, N.: Entwicklung der Abgasemissionen des Personenwagenverkehrs in der BRD von 1970 bis 2010, ATZ 92 (1990) 4
[1.3] Barske, H.: Entwicklungspotential des Kraftstoffverbrauchs von Antriebskonzepten mit Verbrennungsmotor, 3. Aachener Kolloquium Fahrzeug und Motorentechnik, 1991
[1.4] Piech, F.: 3 l/100 km im Jahr 2000, Symposium Energieverbrauch im Straßenverkehr, Wien, 1991, VDI-Verlag Nr. 158
[1.5] N.N.: VDA-Jahresbericht. Auto 89/90, Sept. 1990
[1.6] N.N.: Automotive fuel economy. How far should we go?, National Research Council, National Academy Press, Washington, DC, 1992
[1.7] Bleviss, L.D.: The new Oil Crisis and Fuel Economy Technologies, Quorum Books, Greenwood Publishing, New York, 1988

Chapter 2

[2.1] Schäfer, F.: Thermodynamische Untersuchung der Reaktion von Methanol-Luftgemischen unter der Wirkung von Wasserstoffzusatz, VDI Fortschritt-Berichte, Reihe 6, Nr. 120, 1983
[2.2] Kleinschmidt, W.: Untersuchung des Arbeitsprozesses und der NO-, NO_2- und CO-Bildung in Ottomotoren, Diss. TH Aachen, 1974
[2.3] Bach, W.; Müller, C.F.: Gefahr für unser Klima, Wege aus der CO_2-Bedrohung durch sinnvollen Energieeinsatz, 1982
[2.4] Bach, W.: Klimaveränderung durch Energiewachstum, BWK 31(1979) 2
[2.5] Welsh, D.: Weitere Anzeichen für den Treibhauseffekt, Energie, Jhg. 42, Nr. 4, 1990, S. 54–55
[2.6] N.N.: Bundesgesetzblatt Teil 1, Neufassung der Straßenverkehrszulassungsordnung, Nr. 49, 1988, S. 2016–2017, Bonn
[2.7] N.N.: Bosch, Technische Unterrichtung, Abgastechnik für Ottomotoren, Stuttgart
[2.8] Homann, K.H.: Kohlenwasserstoff-Ruß-Bildungsmechanismus in Flammen, FVV-Kolloquium Ursachen der Kohlenwasserstoffemission von Verbrennungsmotoren, 1977
[2.9] N.N.: California Air Pollution Control Laws, 1989
[2.10] Hardenberg, H.; Albrecht, H.: Grenzen der Rußmassenbestimmung aus optischen Transmissionsmessungen, MTZ 48 (1987) 2
[2.11] Homann, H.S.: Conversion Factors among Smoke Measurements, SAE 850267
[2.12] Förster, H.J.: Entwicklungsreserven des Verbrennungsmotors zur Schonung von Energie und Umwelt-Teil 1, ATZ 93 (1991) 5
[2.13] N.N.: Dritter Bericht der Enquete-Kommission zum Thema Schutz der Erde, Drucksache 11/8030, Bonn, 1988
[2.14] N.N.: VDI-Nachrichten, Juli 1991
[2.15] N.N.: BMFT Ozonforschungsprogramm, Dez. 1988
[2.16] Kümmel, R.; Papp, S.: Umweltchemie, Deutscher Verlag für Grundstoffindustrie, Leipzig, 1988
[2.17] Moussiopoulos, N.; Oehler, W.; Zellner, K.: Kraftfahrzeugemissionen und Ozonbildung, Springer, Berlin Heidelberg New York Tokyo, 1989
[2.18] Metz, N.: Ozon-Mittel zum Zweck?, Automobilrevue Schweiz, Nr. 29, 1991
[2.19] Pischinger, F.: Vorlesungsmanuskript, TH-Aachen
[2.20] Bach, W.: Gefahr für unser Klima. Wege aus der CO_2-Bedrohung durch sinnvollen Energieeinsatz, C.F. Müller, Karlsruhe, 1982
[2.21] N.N.: UPI-Studie, Greenpeace, 1989
[2.22] N.N.: Enquete-Komission, Vorsorge zum Schutz der Erdatmosphäre, Bonn, 1989
[2.23] Grießhammer, R.; Hey, C.; Hennicke, P.: Ozonloch und Treibhauseffekt, Rowohlt, 1990
[2.24] Heck, H. D.: Keine Angst vor CO_2, Bild der Wissenschaft 6 (1989)
[2.25] Ingersoll, A.P.: Die Atmosphäre, Spektrum der Wissenschaft, Heidelberg, 1983
[2.26] Revell, R.: Weltklima: Wärmer und feuchter durch Kohlendioxid
[2.27] Kuhler, M.; Kraft, I.; Klingenberg, H.; Schürmann, D.: Natürliche und anthropogene Emissionen, Automobil-Industrie, 2 (1985)

[2.28] N.N.: Der CO_2-Anstieg in der Atmosphäre – ein weltweites Problem, Umwelt-Bundesamt, Berlin, 1979
[2.29] Zimmermeyer, G.; Esser, F.: CO_2 und Treibhauseffekt, Energiewirtschaftliche Tagesfragen, Heft 9, 1989
[2.30] Kerner, D.; Kerner, I.: Der Klima Report, Kipenheuer & Witsch, Köln, 1990
[2.31] Günther, J.: Bestandsaufnahme in Sachen Ozonloch und Treibhauseffekt, Umweltmagazin, Okt. 89
[2.32] Feister, U.: Ozon – Sonnenbrille der Erde, Teubner Verlag, Leipzig, 1990
[2.33] Schneider, H.: Auto und Umwelt-Perspektiven für das Jahr 2000, ATZ 93 (1991) 1
[2.34] Lavoie, G.A.; Heywood, J.B.; Keck, J.C.: Experimental and Theoretical Study of Nitric Oxide Formation in Internal Combustion Engines, Combustion Science and Technology, Vol. 1, 1970
[2.35] Hoffmannn, R.W.: Aufklärung von Reaktionsmechanismen, Georg Thieme Verlag, Stuttgart, 1976
[2.36] N.N: Gefahrstoffe im Hochschulbereich, Verlag Weinmann, 1992
[2.37] Pischinger, F.: Die Position des Dieselmotors als umweltverträglicher Antrieb, Aachen, 1992
[2.38] Oberdörster, G.; Yu, C.P.: The carcinogetic potential of inhaled Diesel exhaust: A particle effect?, Journal Aerosol Science, Vol. 21, Suppl. 1, 1990
[2.39] N.N.: Auswirkungen von Dieselmotorenabgasen auf die Gesundheit, (Vorabdruck), GSF-For-schungszentrum für Umwelt und Gesundheit, München, 1992
[2.40] Roth, L.; Daunderer, M.: Sicherheitsdaten, MAK-Werte, Ecomed, Landsberg, 1990
[2.41] Bolin, B.; Döös, B.R.; Jäger, J.; Warrick, R.A.: The Greenhous Effekt, Climatic Change and Ecosystems. SCOPE-Report Nr. 29, John Wiley and Sons, Chichester, 1986
[2.42] Schmidt, H.: Reduzierung der Kohlenwasserstoff-Rohemissionen eines Ottomotors beim Kaltstart und bei der instationären Kaltabfahrt, Diss. 1989, TH Braunschweig
[2.43] Pischinger, F.; Schulte, H.; Jansen, J.: Grundlagen und Entwicklungslinien der dieselmotorischen Verbrennung, VDI Berichte, Nr. 714, 1988
[2.44] N.N.: Rußpartikel, FVV-Vorhaben Nr. 261, Abschlußbericht, 1980
[2.45] N.N.: Gefährliche Stoffe, Schriftenreihe der Bundesanstalt für Arbeitsschutz, Bonn, 1985

Chapter 3

[3.1] Bender, K.H.; Ederer, G.; Frerk, J.; Kramer, F.: Der neue BMW Vierzylinder Vierventilmotor, MTZ 50 (1989) 9
[3.2] Miculic, A.; Quissek, F.; Fraidl, G.K.: Sequentielle Einspritzstrategien für verbrauchsoptimierte Ottomotorenkonzepte, MTZ 51 (1990) 7/8
[3.3] N.N.: Bosch Technische Unterrichtungen, Abgastechnik für Ottomotoren, Stuttgart
[3.4] Grohn, M.; Modrich, W.: Nockenwellenversteller von Mercedes-Benz, Automobilrevue Nr. 45/2, Nov. 1989
[3.5] Gruden, D.; Brackert, T.F.; Höchsmann, G.: Motorinterne Maßnahmen zur Minderung der Abgas-Emissionen, VDI-Berichte Nr. 531
[3.6] Basshuysen, R.v.; Stock, D.; Bauder, R.: Audi Turbodieselmotor mit Direkteinspritzung, Teil 3, MTZ 50 (1989)12
[3.7] Larrie, A.; Richeson, W.E.; Erickson, F.L: Performance Evaluation of a Camless Engine Using Valve Actuators with Programmable Timing, SAE 910450, 1991
[3.8] Hiemesch, O.; Lonkai, G.; Scheukermayer, G.: Das BMW Abgasreinigungskonzept für Dieselmodelle, MTZ 51 (1990) 2
[3.9] Pischinger, F.; Schulte, H.; Jansen, J.: Grundlagen und Entwicklungslinien der dieselmotorischen Brennverfahren, VDI Berichte Nr. 714, 1988
[3.10] Abthoff, J.; Fortnagel, M.; Krämer, M.: Das direkteinspritzende Brennverfahren in seiner Eignung als Antrieb für Personenkraftwagen, MTZ 49 (1988) 9
[3.11] Henning, H.: Der Mercedes-Benz "Diesel 89" – Eine Alternative zum geregelten Katalysator, ATZ 91 (1989) 4
[3.12] Emmenthal, K.D.; Grabe, H.J.; Oppermann, W.; Schäpertöns, H.: Motor mit Benzin-Direkteinspritzung und Verdampfungskühlung für das VW Forschungsauto IRVW-Futura, MTZ 50 (1989) 9
[3.13] Seifert, U.: Entwicklungsmöglichkeiten des Dieselmotors, Maschinenwelt Elektrotechnik, Jg.44 Heft 6/7, 1989
[3.14] N.N.: Rußminderung bei Wirbelkammer-Dieselmotoren ohne Sekundärmaßnahmen, Adam Opel AG, Rüsselsheim, 1990
[3.15] Ebbinghaus, W.; Müller, E.; Neyer, D.: Der neue 1, 9-Liter-Dieselmotor von VW, MTZ 50 (1989) 12
[3.16] Basshuysen, R.v.; Stock, D.; Bauder, R.: Audi Turbodieselmotor mit Direkteinspritzung, Teil 1, MTZ 50 (1989) 10
[3.17] Basshuysen, R.v.; Steinwart, J.; Stähle, H.; Bauder, A.: Audi Turbodieselmotor mit Direkteinspritzung, Teil 2, MTZ 50 (1989) 12
[3.18] Basshuysen, R.v.; Stock, D.; Bauder, R.: Audi Turbodieselmotor mit Direkteinspritzung, Teil 3, MTZ 51 (1990) 1
[3.19] Basshuysen, R.v.; Kuipers, G.; Hollerweger, H.: Akustik des Audi 100 mit direkteinspritzendem Turbo-Dieselmotor, ATZ 92 (1990) 1
[3.20] Demel, H.; Stock, D.; Bauder, R.: Audi Turbodieselmotor mit Direkteinspritzung – schadstoffarm nach Anlage 23, MTZ 52 (1991) 9

[3.21] Pischinger, F.: Entwicklungsrichtungen in der Motorentechnik für 2001, 12. Internationales Wiener Motorensymposium, Wien, 1991

[3.22] Bürgler, L.; Herzog, P.; Zelenka, P.: Entwicklungsfortschritte bei PKW-Dieselmotoren zur Erfüllung zukünftiger Emissionanforderungen, 12. Internationales Motorensymposium, Wien, 1991

[3.23] Schneider, H.: Auto und Umwelt, Perspektiven für das Jahr 2000, ATZ 93 (1991) Teil 1

[3.24] Stojek, D.; Stiworok, A.: Valve timing with variable overlap, ISATA 1983, Vol., Paper 82071

[3.25] Schäfer, F.; Hanibal, W.: Die Konstruktion einer Nockenwellenverstelleinrichtung an einem Fünfzylinder Fünfventil-Motor, unveröffentlicher Statusbericht Fa. Audi, Neckarsulm, 1989

[3.26] Fersen, v.O.: Sauber- und Sparmotor von Honda, Automobil Revue, Nr. 34/15, 1991

[3.27] Kreuter, P; Gand, B.; Bick, W.: Beeinflußbarkeit des Teillastverhaltens von Ottomotoren durch das Verdichtungsverhältnis bei unterschiedlichen Hub-Bohrungs-Verhältnissen, 2. Aachener Kolloquium Fahrzeug- und Motorentechnik, 1989

[3.28] Gand, B.: Einfluß des Hub-Bohrungs-Verhältnisses auf den Prozeßverlauf des Ottomotors, Diss. RWTH Aachen,1986

[3.29] Klein, H.: Aufwendiges Magermotor-Management, Automobil Revue Nr. 45/3 Nov. 1988

[3.30] N.N.: Automotive fuel economy. How far should we go?, National Academy Press, Washington, D.C., 1992

[3.31] Müller, E.; Mutke, H.; Neyer, D.: Der 1, 9-l Dieselmotor mit Oxidationskatalysator für den VW-Passat, MTZ 52 (1991)10

[3.32] Brandstetter, W.; Lawrence, P.C.; Hansen, J.: Die Dieselmotoren-Familie mit 1, 8 l von Ford, MTZ 52 (1991) 9

[3.33] Rhode, W.; Gökesme, S.; Liang, J.R.; Schmitt, J.L.: Der neue direkteinspritzende Dieselmotor von Volkswagen, 3. Aachener Kolloquium Fahrzeug- und Motorentechnik, 1991

[3.34] Tippelman, G.: A New Method of Investigation of Swirl Ports, SAE 770404 (1977)

[3.35] Pischinger, F.: Gedanken über den Automobilmotor von morgen, Vortrag VW-AG, Juli 1990

[3.36] Bick, W.: Einflüsse geometrischer Grunddaten auf den Arbeitsprozßß des Ottomotors bei verschiedenen Hub-Bohrungs-Verhältnissen, Diss. RWTH Aachen, 1990

[3.37] Bonse, B.: Einspritzausrüstung für zukünftige PKW-Dieselmotoren, Haus der Technik, Essen, 1991

[3.38] Ohm, I.; Ahn, H.; Lee, W,; Park, S.; Lee, D.: Development of HMC Axially Stratified Lean Combustion Engine, SAE 930879, 1993

[3.39] Schatz, O.: Latentwärmespeicher für Kaltstartverbesserung von Kraftfahrzeugen, BWK Nr. 6, 1991

[3.40] Gehringer, B.: Theoretische und praktische Entwicklung einer variablen Ventilsteuerung auf elektronisch-hydraulischer Basis; VDI-Bericht Nr. 95, Reihe 12, 1989

[3.41] Hockel, K.L.; Langen, P.; Mallog, J.: Abgas-Emissionsreduzierung – eine Herausforderung für die Automobilindustrie, MTZ 53 (1992) 7/8

[3.42] Fortnagel, M.; Moser, P.: Die Mercedes-Benz Dieselmotorenbaureihe für Personenkraftwagen mit Abgasrückführung und Oxidationskatalysator, MTZ 53 (1992) 1

[3.43] N.N.: Federal Certification Test Results for 1991 Model Year, EPA, Washington, D.C., 1991

[3.44] Pischinger, F.: Die Position des PKW-Dieselmotors als umweltverträglicher Antrieb, Studie, 1992

[3.45] Horie, K; Nishizawa, K.: Development of a Four-Valve Lean Burn Engine with VTEC-Mechanism, 13. Wiener Motorensymposium, 1992

[3.46] Ando, H.: Development of the MVV Engine Employing a Novel Lean Burn Concept, Barrel Stratification, 13. Wiener Motorensymposium, 1992

[3.47] Katoh, K; Iguchi, S.; Okano, H.: Toyota Lean Burn Engine–Recent Development, 13. Wiener Motorensymposium, 1992

[3.48] Abthoff, J.; Hüttebräucker, D.; Zahn, W.; Bockel, H.: Die neuen Vierventil-Ottomotoren für die mittlere Baureihe von Mercedes-Benz – Verbrennungs- und abgasseitige Entwicklung der Vierzylindermotoren, MTZ 53 (1992) 11

[3.49] Moser, W. Lange, J, Schürz, W.: Einfluß der Gemischaufbereitungsqualität von Einspritzventilen auf den stationären und instationären Motorbetrieb, 13. Wiener Motorensymposium, 1992

[3.50] Quang-Hue Vo; Oehling, K.H.: Untersuchungen an hydraulischen variablen Ventilsteuerungen, MTZ 52 (1991) 12

[3.51] Mayr, B., Hofmann, R., Hartig, F., Hockel, K.: Möglichkeiten der Weiterentwicklung am Ottomotor zur Wirkungsgradverbesserung, ATZ 81 (1979) 6

[3.52] Hara, S., Kumagai, K., Matsumoto, Y.: Application of a Valve Lift and Timing Control System to an Automotive Engine, SAE 890681, 1989

[3.53] Urata, J.; Umiyama, H.; Shimizu, K; Fujiyoshi, Y; Sono, H.; Fukuo, K.: A Study of Vehicle Equipped with Non-Throttling S.I. Engine with Early Intake Valve Closing Mechanism, SAE 930820, 1993

[3.54] Hatano, K.; Iida, K.; Higashi, H.; Murata, S.: Development of a New Multi-Mode Variable Valve Timing Engine, SAE 930878, 1993

[3.55] Inoue, T.; Matsushita, S.; Nakanishi, K.; Okano, H.: Toyota Lean Combustion System- The Third Generation System, SAE 930873, 1993

[3.56] Walzer, Adamis, Heinrich, Schumacher: Variable Steuerzeiten und variable Verdichtung beim Ottomotor, MTZ 47 (1986) 1
[3.57] Bozung, A. G.; Fleischer, F.: Sercice experiences with different mothods of emission reduction on MAN B+W 4-stroke Diesel and Dual Fuel engines, CIMAC Conference April 1991, Paper 130(D)
[3.58] Nagar, T; Kawahamd, M: Study into Reduction of NO_x Emission in Medium-Speed Diesel Engines, Technical Papers ISME, Kohi '90
[3.59] Wünsche, P.; Wojik, K.: AVL-LEADER, Die neue KKW Dieselgeneration konstruiert für niedrige Emissionen, 14. Int. Wiener Motorensymposium 5, 1993
[3.60] Fraidl, G.K; Quissek, F.; Carstensen, H.: Verbrauchsoptimierte Ottomotorenkonzepte für zukünftige Emissionsszenarien, 13. Int. Wiener Motorensymposium, VDI-Fortschrittberichte, Reihe 12, Nr. 167, S. 35 ff.
[3.61] N.N.: New "Mivec" System. High Efficiency at all Engine Speeds for more Power and better Milage, News from Mitsubishi Motors, Sep. 1992
[3.62] Basshuysen, R.v.: Zylinderabschaltung und Ausblenden einzelner Arbeitszyklen zur Kraftstoffersparnis und Schadstoffminderung, 14. Int. Wiener Motorensymposium 5, 1993
[3.63] Wiedemann, B.; Millmann, M.; Scher, U.: ko-Polo-Antriebskonzept, VDI Berichte 714, 1988
[3.64] Lenz, H.P.: Gemischbildung bei Ottomotoren. Die Verbrennungskraftmaschine, Band 6, Springer, Wien New York, 1990
[3.65] Anisits, F.; Hiemesch, O.; Dabelstein, W.; Cooke, J.; Mariott, M.: Der Kraftstoffeinfluß auf die Abgasemissionen von PKW-Wirbelkammermotoren, MTZ 52 (1991) 5
[3.66] Endres, H.; Krebs, R.: Stand und Entwicklungstendenzen bei Fahrzeug Ottomotoren, FVV-Motorentechnik
[3.67] Gruden, D.: Die Motorenentwicklung im Spiegel der 10. Internationalen Wiener Motorensymposien, VDI-Fortschritt-Berichte, April 1989
[3.68] Moser, F.X.: Kriterien und Potential der Vier-Ventil-Technik bei Nutzfahrzeug-Dieselmotoren, MTZ 50 (1989) 6
[3.69] Hassel, D.; Weber, N.: Ermittlung des Abgasemissionsverhaltens von PKW in der Bundesrepublik Deutschland im Bezugsjahr 1988 (Zwischenbericht), Umweltbundesamt Texte 21/91, Berlin, 1991
[3.70] Pucher, E.: Überprüfung von im Verkehr befindlichen Katalysatorfahrzeugen auf der Basis einer Luftverhältnisbestimmung aus dem Motorenabgas, VDI Fortschrittsberichte, Reihe 12, Nr. 121, 1989
[3.71] Sander, R.: Hält der Kat? Vierjahresbilanz: 20 Autos im Dauertest, mot, Heft 2, 1991
[3.72] Sauer, H.: Kat matt, Umwelt/Katalysator-Prüfung, Auto Motor Sport, Heft 6, 1990
[3.73] Brosthaus, J.: Emissionen von Kraftfahrzeugen, TÜV Rheinland, Seminar "Verkehrsbedingte Immissionen in den Städten", Haus der Technik, Essen
[3.74] N.N.: Feldüberwachung der Abgasemissionen in der Bundesrepublik Deutschland, F+E Vorhaben, RWTÜV Essen
[3.75] Pischinger, F.: Vorlesungsmanuskript TH Aachen, 1988
[3.76] Cichocki, R.; Schweizer, F.: Entwicklungspotential des schnellaufenden PKW-Dieselmotors mit direkter Einspritzung, AVL-Vortrag Haus der Technik, 1992
[3.77] Anisits, F.; Reibold, G.: Möglichkeiten zur Partikelbegrenzung bei PKW-Dieselmotoren, Automobil-Industrie, 6/89
[3.78] Schäpertöns, H.; Emmenthal, K.D.; Grabe, J.; Oppermann, W.: Das Motorenkonzept des VW- Forschungsmotors, 2. Aachener Kolloquium Fahrzeug- und Motorentechnik, 1989, S. 51
[3.79] Schulte, H.; Dürnholz, M.; Wübbeke, K.: Die Rolle des Einspritzsystems bei der Erfüllung zukünftiger Schadstoffemissionsgrenzwerte, 2. Aachener Kolloquium Fahrzeug- und Motortechnik, 1989, S. 303
[3.80] Menne, J.; Königs, M.: Magerkonzepte – Eine Alternative zum Dreiwegekatalysator, MTZ 49 (1988) 10
[3.81] Fraidl, K.G.; Quissek, F.; Carstensen, H.: Verbrauchsoptimierte Ottomotorkonzepte für zukünftige Emissionsszenarien, MTZ 54 (1993) 4

Chapter 4

[4.1] Knoll, R.: Kraftstoffeinbringung sowie experimentelle und rechnerische Methoden zur Ladungswechselverbesserung, AVL,1991
[4.2] N.N.: Ost-West-Forum Verkehr, Der Zweitakt-Motor im Kraftfahrzeug, Kolloquium TU Berlin, 1990
[4.3] Spiegel, L.; Spicher; U.; Bäcker, H. Lingen, A.: Untersuchungen zum Spülverlauf und zur Verbrennung im 2-Takt-Motor, Haus der Technik, Feb. 1993
[4.4] Duret, P.: The key points for the development of an automotive spark ignition two-stroke engine, IMechE, 1992
[4.5] Ando, H.: Assessment of the Feasibility of the Two-Stroke-Cycle-Engine with Poppet Valves, Haus der Technik, Feb. 1993
[4.6] Emmenthal, K.D; Schäpertöns, H; Oppermann, W.: Emissionen des modernen Zweitakt-Motors im Vergleich zum Viertakt-Motor, VDI-Tagung Fahrzeugmotoren im Vergleich, Dresden, 1993

[4.7] Behrens, M.: Schlußfolgerungen aus Entwicklungsarbeiten der TH Zwickau an Zweitakt-Motoren, VDI-Tagung Fahrzeugmotoren im Vergleich, Dresden, 1993
[4.8] Mallog, J.; Theissen, M.; Heck, E.: 2-Takt-Motor – Antriebskonzept der Zukunft?, VDI-Tagung Fahrzeugmotoren im Vergleich, Dresden, 1993
[4.9] Karl, G.; Mikulic, L.; Schommers, J.; Frieß, W: Abgas, Verbrauch, Leistung, Komfort – wo liegen die Chancen und Risiken für einen modernen PKW-Zweitakt-Dieselmotor?, VDI-Tagung Fahrzeugmotoren im Vergleich, Dresden, 1993
[4.10] Fraidl, G.K.; Knoll, R.; Hazeu, H.P.: Direkte Gemischeinblasung am 2-Takt-Ottomotor, VDD-Tagung Fahrzeugmotoren im Vergleich, Dresden, 1993
[4.11] Pfeifer, U.; Zarske, S: Der Beitrag des Schmierungssystems zur Schadstoffemission von Zweitaktmotoren, VDI-Tagung Fahrzeugmotoren im Vergleich, Dresden, 1993

Chapter 5

[5.1] Pischinger, F.: Verbrennungsmotoren, Vorlesungsumdruck, RWTH Aachen
[5.2] Koberstein, E.: Abgaskatalysatoren, Bauarten Funktion Verfügbarkeit
[5.3] N.N: Technische Information Fa. Pierburg, Neuß
[5.4] N.N: Emissionsminderung Automobilabgase – Ottomotoren, VDI-Verlag, 1987
[5.5] Öser, P.; Brandstetter, W.: Grundlagen zur Abgasreinigung von Ottomotoren mit Katalysatortechnik, MTZ 45 (1984) 5
[5.6] Engler, B.; Brand, R.: Katalysatoren fur den Umweltschutz, Dokumentation Degussa, Hanau, 1990
[5.7] Oblander, K.; Abthoff, J.; Schuster, H. D.: Der Dreiwege-Katalysator – eine Abgasreinigungstechnologie für Kraftfahrzeuge mit Ottomotor, VDI-Bericht 531
[5.8] Brill, U.; Heuber, U.: Werkstoffe für Metallträger von Automobil-Abgaskatalysatoren, MTZ 49 (1988) 9
[5.9] Nonnenmann, M: Metallträger für Abgaskatalysatoren in Kraftfahrzeugen, MTZ 45 (1984) 12
[5.10] N.N.: Optimierte Katalysatortechnik, MTZ (1991) 10
[5.11] Förster, H.J.: Entwicklungsreserven des Verbrennungsmotors, ATZ 93 (1991) 6
[5.12] N.N.: Presseinformation Daimler Benz, 12-Zylindermotor, 1991
[5.13] Schneider, H.: Auto und Umwelt, Perspektiven für das Jahr 2000, ATZ 93 (1991) 1
[5.14] N.N.: Abgastechnik für Ottomotoren, Technische Unterrichtung Bosch, Stuttgart, 1989
[5.15] Zelenka, P.: Einsatz von Oxidationskatalysatoren an Dieselmotoren für die Erfüllung zukünftiger Abgasemissionsgrenzwerte, Vortrag TU Berlin, 1991
[5.16] Pischinger, F.; Schulte, H.; Jansen, J.: Grundlagen und Entwicklungslinien der dieselmotorischen Brennverfahren, VDI-Berichte 714, 1988
[5.17] Muller, E.: Der VW Kat-Diesel mit leichter Aufladung, 2. Aachener Kolloquium Fahrzeug und Motortechnik, 1989
[5.18] Fellhofer, H.: Möglichkeiten zur Verminderung von Emissionen direkteinspritzender Dieselmotoren, VDI-Berichte,1988
[5.19] N.N.: Zweiter Fortschrittbericht zum Stand der Entwicklung von Partikelfiltern für Dieselmotoren, VDA, 1988
[5.20] Handorf, H.; Grunwald, B.: Thermische Regeneration von Partikelfiltern durch temperatursteigernde motorische Zusatzmaßnahmen, VDI-Berichte 885, 1991
[5.21] Blümel, H.: Rußfilter Großversuch, Schadstoffbelastungen Konzept Zwischenergebnisse, VDI-Berichte 885, 1991
[5.22] Hardenberg, H.: Das keramische Wickelfilter mit katalytischer Regeneration für DB-Nutzfahrzeuge, Verkehr und Technik, Heft 6
[5.23] Huber, G.; Obländer, K.: Recycling am Mercedes, Teil 2, ATZ 93 (1991) 1
[5.24] N.N.: Interatom Blähmatte, Lagerungssystem für katalytische Konverter, 1989
[5.25] Willenbockel, O.; Friedrich, A.; Arnold, G.: Der neue 3 Liter Vierventilmotor von Opel, MTZ 50 (1989) 10
[5.26] Plapp, G.; Glöckler, O.; Schnaibel, E.: Gemischregelung für optimalen Betrieb eines Dreiwege Katalysators, 3. Aachener Kolloquium Fahrzeug und Motorentechnik, 1991
[5.27] N.N.: Abgas Schnüffler, mot-Technik 113, 1991
[5.28] König, A.: Abgasnachbehandlung beim Dieselmotor, Haus der Technik, Essen, 1991
[5.29] N.N.: Neuer Rußfilter von SHW, MTZ 52 (1991) 11
[5.30] Härkönen, M.; Malvile, P.: Optimisation of Metallic TWC Behaviour and Precious Metal Costs, SAE 920395
[5.31] Ma, T.; Collings, N.; Hands, T.: Exhaust Gas Ignition (EGI) – A New Concept for Rapid Light Off of Automotive Exhaust Catalytic Converters, SAE 920400
[5.32] Herzog, P; Bürgler, L; Winkelhofer, E; Zelenka, P; Cartellieri, W.: NO_x-Reduction Strategies for DI Diesel Engines, SAE 920470
[5.33] Puppe, L.: Zeolites-Properties and Practical Applications, Chemie in unserer Zeit, Vol.20, No.4, pp 117–127, 1986

[5.34] Polach, W; Hägele, K.H.: Erste Ergebnisse mit dem elektrostatischen Rußfilter an einem PKW-Dieselmotor, XX Fisita Congress Wien, Mai 1984
[5.35] N.N.: Katalysator Startheizungssystem, MTZ 54 (1993) 1
[5.36] Buch, D.: Der Abgaskatalysator, Aufbau, Funktion und Wirkung, Schriftenreihe Opel AG, Nr. 42, Rüsselsheim, 1984
[5.37] N.N.: Donaldson Rußabbrennfilter, unveröffentlichte Information, 1993
[5.38] Kaiser, W.F.; Maus, W.; Swars, H.; Brück, R.: Optimisation of an Electrically-Heated Catalytic Converter System – Calculations and Application, SAE 930384, 1993
[5.39] Hellmann, K.H.; Piotrowski, G.K.; Schaefer, R.M.: Start Catalyst Systems Employing Heated Catalyst Technology for Control of Emissions from Methanol-Fueled Vehicles, SAE 930382, 1993
[5.40] Hochmuth, J.K.; Burk, P.L.; Tolentino, C.; Mignano, M.J: Hydrocarbon Traps for Controlling Cold-Start Emissions, SAE 930739, 1993
[5.41] Minami, T.; Nagase, T.: Exhaust Gas Purification Device in Variable Combination of Absorbent and Catalyst According to Gas Temperature, EP 424966, 1990
[5.42] Engler, B.H.; Lindner, D.; Lox, E.S.; Ostgathe,K.; Schäfer-Sindlinger, A. Müller, W.: Reduction of Exhaust Gas Emissions by Using Hydrocarbon Adsorber Systems, SAE 930738, 1993
[5.43] Kojetin, P.; Janezich, F.; Sura, S.; Tuma, D.: Production Experience of a Ceramic Wall Flow Electric Regeneration Diesel Particulate Trap, SAE 930129, 1993
[5.44] Schweizer, W.: Praxis Katalysator Autos, Krafthand Verlag, Darmstadt, 1990
[5.45] Richochi, R.: Entwicklungspotential des schnellaufenden PKW-Dieselmotors mit direkter Einspritzung, Haus der Technik, Essen, 1991
[5.46] Schäfer, F.; Vogel, H.; Heinze, T.: Dieselkatalysator für Nutzfahrzeuge unter besonderer Berücksichtigung des Einsatzes im innerstädtischen Bereich, ATZ (1993) 12
[5.47] Müller, E.; Wiedemann, B.: Dieselpartikelfiltersystem mit additivgestützter Regeneration, ATZ (1991)
[5.48] Schmidt, S.; Schäfer, F.: Entwicklung eines Zweirad-Katalysators, VDI-Bericht Nr. 1025, 1993
[5.49] Anisits, F.; Reibold, G.: Möglichkeiten zur Partikelbegrenzung bei PKW-Dieselmotoren, Automobilindustrie 6/89
[5.50] Pelters, S.; Kaiser, W.; Maus, W.: The Development and Application of a Metal Supported Catalyst for Porsche's 911 Carrera 4, SAE 890488
[5.51] Maus, W.; Bode, H.; Reck, A.: Neuentwicklung metallischer Katalysatorträger für Automobile – Einfluß der Katalysatorträger auf Motor und Abgas, Technische Akademie Wuppertal, 1989
[5.52] Nonnenmann, M.: Neue Metallträger für Abgaskatalysatoren mit erhöhter Aktivität und innerem Strömungsausgleich, ATZ 91 (1989) 4

Chapter 6

[6.1] Hirao, O.; Pefley, R.K.: Present and Future Automotive Fuels, J. Wiley u. Sons, New York, 1988
[6.2] Dabelstein, W.; Reglitzky, A.; Reders; Lucht: 100 Jahre Kraftstoffe für den Straßenverkehr, Shell AG, Hamburg, 1987
[6.3] Nierhauve, B.: Heutige und zukünftige Kraftstoffe für Ottomotoren, Techn. Akadem. Wuppertal, 1989
[6.4] Gruden, D.; Zeller, D.; Höchsmann, G.: Einfluß der Kraftstoffzusammensetzung auf Betriebsverhalten und Abgasemissionen des Ottomotors, 8. Internationales Motorensymposium, 1987, Wien
[6.5] Höchsmann, G.; Zeller, D.: Entstehung und Minderung der Benzolemission aus Ottomotoren, BMFT-Statusseminar, 1986
[6.6] Anisits, F., Hiemesch, O.; Dabelstein, W.; Cooce, J.; Mariott, M.: Der Kraftstoffeinfluß auf die Abgasemissionen von PKW-Wirbelkammermotoren, MTZ (1991) 5
[6.7] Fortnagel, F.: Der Dieselmotor im PKW unter Emissionsaspekten-Möglichkeiten und Probleme, 2. Aachener Kolloquium, 1989
[6.8] Houben; Lepperhoff: Der Kraftstoffeinfluß auf die Partikelemissionen von Dieselmotoren, Beratungsgesellschaft für Mineralöltechnik, 1990
[6.9] Schoonveld, G.A.; Marshall, W.F.: The Total Effect of a Reformulated Gasoline on Vehicle Emissions by Technology, SAE-Paper 910380
[6.10] Berg, W.: Die neue Abgasgesetzgebung der USA, VDI Fortschrittberichte Nr. 150, Reihe 12, 1991
[6.11] N.N: Auf dem Weg zu neuen Energiesystemen, Teil I,II,III, Programmstudie Bundesministerium für Forschung und Technologie, Berlin, 1975
[6.12] Pischinger, F.; Adams, W.: Abgasemissionen bei Verwendung alternativer Kraftstoffe für Kraftfahrzeug-Ottomotoren, VDI-Bericht 531
[6.13] Harrington, I.A.; Shishu, R.C.: A Single-Cylinder Engine Study of the Effects of Fuel Type, Fuel Stoichiometry and Hydrogen-to-Carbon Ratio on CO, NO and HC Exhaust Emissions, SAE-Paper 730476
[6.14] Menrad, H.; Steinke, D.; Wegener, R.: Abgasemissionen mit alternativen Kraftstoffen, MTZ 47 (1986) 2

[6.15] N.N.: Flüssiggas – ein Alternativkraftstoff, RW-TÜV-Essen, Schriftenreihe, 1982
[6.16] Peschka, W.: Flüssiger Wasserstoff als Energieträger, Springer, Wien – New York, 1984
[6.17] Murray, R.G.; Schoeppel, R.J.; Gray, C.L.: The Hydrogen Engine in Perspective, Intersociety Energy Convention Engineering Conference, 1972
[6.18] Kuhlmann, A.; May, H.; Pischinger, F.: Methanol und Wasserstoff Automobilkraftstoffe der Zukunft, Verlag TÜV Rheinland, 1976
[6.19] Binder, K.; Witthalm, G.: Mixture Formation and Combustion in Interaction with the Hydrogen Storage Technology, Proc., 3rd World Hydrogen Energy Conference, Vol 2, Tokyo, 1980
[6.20] Watson, H.C.; Milliken, E.E; Deslandes. J.V.: Efficiency and Emissions of a Hydrogen or Methane Fuelled Spark-Ignition Engine, FISITA, Paris, 1974
[6.21] Finegold, J.G.: The UCLA Hydrogen Car: Design Construction and Performance, SAE 730507, 1973
[6.22] Schäfer, F.: Reaktionskinetische Untersuchungen des Wasserstoff-Benzin Mischbetriebes und deren Auswirkungen auf den Ottomotor, Diss. Universität Kaiserslautern, 1979
[6.23] May, H.; Gwinner, D.: Möglichkeiten der Verbesserung von Abgasemissionen und Energieverbrauch bei Wasserstoff-Benzin Mischbetrieb, FISITA, Hamburg, 1980
[6.24] Kapus, P.; Miculic, L.; Zelenka, P.: Entwicklung eines direkteinspritzenden Methanolmotors für PKW, 11. Intern. Wiener Motorensymposium, 1990.
[6.25] Jordan, W.: Erweiterung des ottomotorischen Betriebsbereichs durch Verwendung extrem magerer Gemische unter Einsatz von Wasserstoff als Zusatzkraftstoff, Diss. Universität Kaiserslautern, 1977
[6.26] Förster, H.: Entwicklungsreserven des Verbrennungsmotors, ATZ 93 (1991) 6
[6.27] N.N.: Mineralöltechnik 5–6, Mai 1989
[6.28] Menrad, H.; Weidman, B.: Rapsölmetylester im Dieselmotor, MTZ 50 (1989) 2
[6.29] Pächter, H.: RME als Dieselkraftstoffersatz, AVL-Tagung Motor und Umwelt, Graz, 1991
[6.30] Fabri, J.; Dabelstein, E.A.; Reglitzky, A.: Chancen alternativer Kraftstoffe unter besonderem Aspekt der Umweltverträglichkeit, Shell Studie, Hamburg, 1990
[6.31] Richter, H.: Einsatz von Rapsöl als Alternativkraftstoff für Dieselmotoren, AVL Tagung Motor und Umwelt, Graz, 1991
[6.32] Brand, K.G.: Kraftstoffadditive – ein Beitrag zur sauberen Umwelt, AVL Sommerakademie, 1991
[6.33] Paramius: Exxon chemical International Marketing, B.V., 1991
[6.34] Menrad, H.; Wegener, R.; Loeck, H.: An LPG-optimized Engine-Vehicle-Design, SAE 852071, 1985
[6.35] Nierhauve, B.: Heutige und zukünftige Kraftstoffe für den Ottomotor, Techn. Akademie Wuppertal, Wuppertal, 1989
[6.36] N.N.: Studie über den Einsatz von H_2 beim Wankelmotor, interner Untersuchungsbericht Fa. Audi, Neckarsulm, 1978
[6.37] Bayerl, W.; Müller, A.; Schäfer, H.V.: Einfluß der Kraftstoffzusammensetzung auf die Kohlen- wasserstoffemissionen von Ottomotoren, MTZ 53 (1992) 12
[6.38] N.N.: Ansatzpunkte und Potentiale zur Minderung des Treibhauseffektes aus Sicht der fossilen Energieträger, DGMK Forschungsbericht 448, 1991
[6.39] N.N.: Erdöl Informationsdienst Nr. 41, 1991
[6.40] Wagner, U.: Spezifische Emissionen bei elektrisch und konventionellen PKW-Antrieben, VDI-Bericht 985, 1992
[6.41] Bayerl, W; Müller, A; Schäfer, H.V.: Einfluß der Kraftstoffzusammensetzung auf die Kohlenwasserstoffemissionen von Ottomotoren, MTZ 53 (1992) 12
[6.42] Syassen, O.: Chancen und Problematik nachwachsender Kraftstoffe, MTZ 53 (1992), 11 und 12
[6.43] Völz, M.; Klüver, D.; Meins, H.J.; Stempel, G.: Entwicklungsschwerpunkte bei Schmierstoffen für moderne Nutzfahrzeugmotoren, MTZ 53 (1992) 1
[6.44] N.N.: Motorölgeneration für geeignete Ansprüche, Automobil Revue Nr. 7/18.2, Bern, 1993
[6.45] Nikanjam, M.: Development of the First CARB Certified California Alternative Diesel Fuel, SAE 930728, 1993
[6.46] Förster, H.J.: Der ideale Kraftstoff aus der Sicht des Fahrzeugingenieurs – Teil 1, ATZ 84 (1992) 4; Teil 2, ATZ 84 (1982) 5
[6.47] Huynh, N.; Krickelberg, T.; Richter, H.; Schulz, H.: Erfahrungen im Hause Porsche mit Methanolkraftstoffen, VDI-Bericht nr. 1020, Düsseldorf, 1992
[6.48] Seiffert, U.; Held, W.: Alternative Kraftstoffe: Chancen und Aufgaben, Chem.-Ing.-Technik 53 (1981) 2
[6.49] Knaak, M.; Reglitzky, A.: Untersuchungen über den Einfluß von Motorölen auf die Wirksamkeit von Abgaskatalysatoren, Shell Technischer Dienst, Hamburg, 1988
[6.50] Pischinger, F.; Hilger, U.; Jain, G.; Bernhardt, W.; Heinrich, K.; Weidmann, K.: Konzept eines 1, 9 l-DI-Methanolmotors für den Einsatz im PKW, 11. Internationales Wiener Motorensymposium, Wien, April 1990,

Chapter 7

[7.1] Seifried, D.: Gute Argumente – Verkehr, Beck-Verlag, München, 1991
[7.2] Dursbeck, F.: Ermittlung der CO_2-Emissionen für PKW mit Otto- und Dieselmotoren, Umweltbundesamt, Berlin, 1989

[7.3] Abthoff, J.; Noller, C.; Schuster, H.: Möglichkeiten zur Reduzierung der Schadstoffe von Ottomotoren, Fachbibliothek Daimler-Benz, 1983
[7.4] Heitland, H.; Hiller, H.; Hoffman, H.J.: Einfluß des zukünftigen PKW-Verkehrs auf die CO_2-Emissionen, MTZ 51 (1990)
[7.5] Fabri, I.; Dabelstein, W.; Reglitzky, A.: Chancen alternativer Kraftstoffe unter dem Aspekt der Umweltverträglichkeit, Shell, Technischer Dienst, Hamburg, 1990
[7.6] Sporkmann, B.: Elektrofahrzeuge als Luftschadstoffbremse, Energiewirtschaftliche Tagesfragen, 1990, Heft 6
[7.7] Höchsmann, G.; Gruden, D.: Alkohole als alternative Kraftstoffe, Automobil-Industrie, 1/1989
[7.8] N.N.: Abgasemissionen, Schriftenreihe der Fa. KHD Deutz, Köln, 1989
[7.9] Förster, H.: Entwicklungsreserven des Verbrennungsmotors, ATZ 93 (1991) 6
[7.10] Anisits,F.; Hiemesch, O.; Dabelstein, W.; Cooke, J.; Mariott, M.: Der Kraftstoffeinfluß auf die Abgasemissionen von PKW-Wirbelkammermotoren, MTZ 52 (1991) 5
[7.11] Schneider, H.: Auto und Umwelt, Perspektiven für des Jahr 2000, ATZ 93 (1991)
[7.12] Hassel, D.; Weber, F.J.: Ermittlung des Abgasemissionsverhaltens von PKW in der Bundesrepublik Deutschland im Bezugsjahr 1988, Umweltbundesamt 21/90, Berlin, 1990
[7.13] Pischinger, F.: Studie: Die Position des Dieselmotors als umweltverträglicher Antrieb, 1992
[7.14] N.N.: Ansatzpunkte und Potentiale zur Minderung des Treibhauseffektes aus Sicht der fossilen Energieträger, DGMK Forschungsbericht 488, 1991
[7.15] Demel, H.: Wie lange ist die Verbrennungskraftmaschine beim PKW-Antrieb noch sinnvoll?, 14. Wiener Motorensymposium, Wien, 1993

Chapter 8

[8.1] N.N.: Fa. Pierburg, Abgasmeßtechnik, technische Information, Neuß, 1993
[8.2] N.N.: Kommission der europäischen Gemeinschaften, Vorschlag für eine Richtlinie des Rates zur Änderung der Richtlinie 70/220/EWG zur Angleichung der Rechtsvorschriften der Mitgliedsstaaten über Maßnahmen gegen die Verunreinigung der Luft durch Kraftfahrzeugemissionen, (Februar 1990); S. 1–10, 60–70, Brüssel
[8.3] N.N.: Umweltbundesamt, KFZ-Abgas-Emissionen, Grenzwerte Vorschriften Messungen, Berlin, 1989
[8.4] N.N.: Fa. Pierburg: Bedienungsanleitung AMA 2000, Neuß, 1993
[8.5] Pischinger, F.: Vorlesungsmanuskript RWTH Aachen, 1988
[8.6] Dutta, R.: Abgasmessungen an Ottomotoren – Probenahme und Analytik, VDI-Berichte 531, 1984
[8.7] N.N.: Neuer Mehrkomponenten-Gasanalysator für Motorenabgase, MTZ 51 (1990) 4
[8.8] Zelenka, P.: Einsatz von Oxidationskatalysatoren an Dieselmotoren für die Erfüllung zukünftiger Abgasgrenzwerte, Vortrag TU Berlin, 1991
[8.9] Schäfer, F.: Gesetzliche Vorschriften zur Schadstoff- und Verbrauchsbegrenzung bei PKW-Verbrennungsmotoren, MTZ 52 (1991) 7/8
[8.10] N.N.: Bundesgesetzblatt Nr. 49, Teil 1, Bonn, 1988
[8.11] N.N.: California Air Recources Board, Initial Statement of Reasons for Proposed Rulemaking, (April 1989); S. 1–4, 7–8, 12, 31
[8.12] Goldammer, W.; Rummel, R.: Auf dem Weg zum abgasarmen Auto, (Dezember 1984); S. III, 1–6, 42–45, 47–51
[8.13] N.N.: Daimler-Benz, Faltblatt zur Abgasgesetzgebung, Stuttgart, 1991
[8.14] Stern, A.C.: University of North Carolina, History of Air Pollution Legislation in the United States, (1982); S. 52–57
[8.15] Berg, W.: Aufwand und Probleme für Gesetzgeber und Automobilindustrie bei der Kontrolle der Schadstoffemissionen von PKW mit Otto- und Dieselmotoren, Diss. TU Braunschweig, 1981
[8.16] N.N.: Umweltbundesamt, Kfz-Abgas-Emissionen, Grenzwerte – Vorschriften – Messungen, (Januar 1989), Berlin
[8.17] N.N.: Faltblatt Emissionsgrenzwerte 1989, VW AG, Wolfsburg
[8.18] N.N.: Environment Agency, Japan, Motor Vehicle Pollution Control in Japan (2nd Revision, 1987); S. 1, 4, 16–24, 32–33, Tokyo
[8.19] Seiffert, U.: Status of German/European Exhaust Emission Legislation, SAE 851211
[8.20] Walsh, M.P.: NO_x Emissions from Road Traffic in Europe – Projections beyond the Year 2000, For Presentation at Workshop on Projections of NO_x Emissions – Oslo, Norway, December 1989
[8.21] N.N.: Luftreinhaltung im Verkehrsbereich, Umwelt-Bundesminister, (März 1990); S. 10,16–21, 28, Bonn
[8.22] N.N.: U.S. EPA, Clean Air Facts, Nr. 3, 5, 9, 10, 15/1989, Washington, D.C.
[8.23] N.N.: Europäisches Umweltbüro EEB, Higlights of the Bigpartisan Senate Clean Air Act Agreement, (Februar 1990); S. 1–2, 6–9
[8.24] N.N.: California Air Resources Board, Low-Emission Vehicles/Clean Fuels and new Gasoline Specifications-Progress report, (Dezember 1989); S. 1–4, 9–10, 13–14, 28, Sacramento
[8.25] N.N.: California Air Resources Board, Technical Support Document (Juni 1989); S. I/50–52, 67–68 , II/18, Sacramento

[8.26] N.N.: Automotive Fuel Economy. How Far Should We Go?, National Research Council, National Academy Press, Washington, DC, 1992

[8.27] N.N.: Ministry of Transport, Japan, MOT NEWS Nr. 37/1989, Nr. 40, 41/1990

[8.28] N.N.: Japan Motor Industrial Federation, Inc., Future of the Japanese Automotive Industry, (Juli 1989); S. 8, 30 ff.

[8.29] N.N.: EG-Richtlinie vom 3. Dezember 1987 zur Änderung der Richtlinie 70/220/EWG über die Angleichung der Rechtvorschriften der Mitgliedsstaaten über Maßnahmen gegen die Verunreinigung der Luft durch Abgase von Kraftfahrzeugmotoren – 88/76/EWG, Brüssel

[8.30] N.N.: EG-Richtlinie vom 18. Juli 1989 zur Änderung der Richtlinie 70/220/EWG zur Angleichung der Rechtsvorschriften der Mitgliedsstaaten über Maßnahmen gegen die Verunreinigung der Luft durch Emissionen von Kraftfahrzeugen hinsichtlich der europäischen Emissionsnormen für Kraftfahrzeuge mit einem Hubraum unter 1, 4 Litern – 89/458/EWG, Brüssel

[8.31] N.N.: Vierter Immissionsschutzbericht der Bundesregierung, Umwelt-Bundesminister, (Juli 1988); S.75, Bonn

[8.32] N.N.: Pressemitteilung: Die Bundesregierung sorgt für endgültigen Durchbruch beim schadstoffarmen Auto, (Juli 1989)

[8.33] N.N.: Schärfere Grenzwerte geben grünes Licht für den Dreiwegekatalysator, VDI-Nachrichten, Nr. 2, 1991

[8.34] N.N.: Bundesgesetzblatt Teil 1, Nr. 49, 1988, Bonn

[8.35] Kemper, G.: Initiative der europäischen Gemeinschaft und der Bundesrepublik zur Minderung der Schadstoffemissionen und des Kraftstoffverbrauchs von Verbrennungsmotoren, AVL-Tagung Motor u. Umwelt, Graz 1991

[8.36] Walther-Borjans, N.: Öko-Steuer – Umweltschutz aus Eigennutz?, Zeitschrift: "fairkehr", 1990

[8.37] N.N.: Institut für Energietechnik und Umweltschutz, TÜV Rheinland, Vorschlag für eine umweltorientierte Kfz-Steuer, (Mai 1990)

[8.38] Plotkin, S.: OTA Perspective, Office of Technology Assessment, U.S. Congress, SAE, 1992

[8.39] N.N: Betriebsanleitung BINOS 100, Leybold AG

[8.40] N.N.: Rußpartikel, FVV Abschlußbericht Nr. 261

[8.41] Klingenberg, H.: Neues Meßsystem zur gleichzeitigen zeitaufgelösten Direktmessung von limitierten und nichtlimitierten Automobilabgaskomponenten, 7. Internationales Wiener Motorensymposium, 1986

Subject index

Acetaldehyde 119
Acetyl acetone 113
Acid rain 12, 14, 15, 17
Acoustic 65
Activation energy 4
Additive 7, 97, 136–138
Advanced idle setting 31
Aerosol 14
Ageing of the catalytic converter 96–98
Air deficiency 8, 89
Air dilution 150
Air-distributing (diesel combustion process) 48, 54, 64
Air envelope around the injection jet 24, 25
Air/fuel ratio 8, 10–12, 20, 23, 24, 27, 88, 93, 96, 143
Air quality 156
Air quantity 11
Alcohol 10, 117, 119, 123, 126, 127, 132
Aldehydes 4, 6, 10, 57, 126, 131, 132
Alternative fuel 37, 47, 122, 124, 126, 132, 136
Annex-23 58
Annex-25 58
Annual exhaust test 163
Anthropogenic emission 14
Anti-knock additive 10
Aromates/aromatic substances 6, 10, 126
Atmosphere 13
Auxiliary chamber 51
Average mileage 1, 2

Backfiring 131
Benzene 6, 10, 118, 119
Biofuel 133
Biogas 123
Biomass 145
Bore-stroke ratio 24, 34
Butane 129

CAFE standard 157, 162
Camshaft adjustment 30
CARB 12, 127, 167, 168
Carbon dioxide 7, 17, 18
Carbon monoxide 4, 6, 8, 11, 89
Carcinogenic 6
Carnox 103
Catalytic converter 1, 10, 27, 38, 86, 88, 89, 102, 105, 106, 128
–, heated 47, 94
Catalytic converter ageing 72
Catalytic converter contamination 96–98, 102, 104
Catalytic converter efficiency 171
Catalytic matrix 92
Categories of engine displacement 175
Ceramic catalytic converter 90
Ceramic foam filter 108
Ceramic monolith filter 108

Ceramic yarn wrap-around filter 108, 114
Certification procedure 164
Cetane number enhancer 138
Cetane rating 120, 121, 138
CFC 15, 16, 171
CH_4 16
Chamber geometry 51
Chamber volume 51
Charge changing process 9
Charge changing work 78, 80
Charge dilution 33
Charge loss 33
Charge movement 21, 24, 26
Charge stratification 11, 24, 44
Charging 63, 65
Charging/filling port 24, 33, 62
Charging state 106, 110
Chemical luminescence/chemiluminescence detector 152, 153
Chemical reaction 3
Chlorofluorocarbon 18
"Clean Air Act" 119, 122, 156
"Clean fuel" 165, 166, 168
Climate 13, 15
Climate mechanism 13
CO emission 9, 26, 37, 38, 48, 51, 57, 86, 87, 119, 122, 125, 128, 133, 134, 138, 142, 163, 164, 177
CO formation 9, 16
CO_2 concentration 18
CO_2 emission 18, 27, 38, 59, 71
Cold start 60, 95, 140
Cold-temperature test procedure 164, 171
Combustion chamber 9, 12, 36
– bowl 57
– shape 21, 23, 24
– surface 21, 34
Combustion duration 40, 138
Combustion enhancer 138, 139
Combustion layout 9, 20, 21
Combustion noise 27, 138
Combustion path 23
Combustion process 12, 13, 48, 144
Combustion speed 28, 40
Combustion temperature 28, 54, 124, 127
Comparison process 20
Component temperature 35
Compression ratio 8, 20–22, 43, 64
Concentration 12
Conditioning 150
Consumption standards 147
Content of aromatic substances 118, 119, 171
Conventional fuel 117, 124
Conventional injection 83
Conversion behavior 95, 96
–, dynamic 96

Conversion behavior, static 96
Conversion efficiency 25, 57, 93, 94
Cooling 35
Copper oxide 113
Cordierite 90
Crank angle 31
Crankcase scavenging system 79
Curb weight 143
Cutoff in overrunning mode 99
CVS method 54, 57, 58, 150
Cylinder cutout 32, 47
Cylinder head 24
Cylinder volume 34
Cylinder-selective 25, 26, 102

Damping mat 93
Dead volume 56
Death risk 7
Demulsification 139
Denox-catalyst 47, 62, 86
Deposit formation 137, 138
Deposition rate 106
Depth-type trap 107
Detergents 138
Deterioration factors 75, 177
DI diesel 61, 122
Diesel additive 139
Diesel combustion process 48
Diesel engine 12, 102
Diesel engine exhaust 7
Diesel engine fuel 7, 8, 144
Diesel fuel 120, 122
Diesel smoke meter/analyzer 154
Diesel soot 6
Dilution tunnel 150, 152, 155
Direct-injection 11, 42, 44, 45, 48, 54–57, 69–71, 73, 74
Distributor-type injection pump 67
Dual-bed catalytic converter 89

Earth atmosphere 14, 15
ECE 15/04 162
ECE regulations 58, 121, 162
ECE test cycle 7, 148
EEC 162, 174
EEC emission regulation 55
EEC regulation 162
Effective compression ratio 43
Effective efficiency 43, 77
Efficiency 21, 34, 35
EGR rate 22, 68, 69
Electrical drive 135, 137, 144
Electrical filter 109
Electronic control of diesel 51, 57, 67, 68
Elementary reaction 3
Eleven-mode test cycle 159, 160
Emission coefficients 143
Emission maps 71–74
Emission measurement methods 150
Emission measuring instruments 150, 152
Emission scatterbands 75, 76
Emission standard 24, 29, 156, 157, 164–168, 170, 172–174, 176–178, 180

Emission standards in the Federal Republic of Germany 162, 163, 178
Emission test cycle 95
Emission testing procedure 147, 150
Energy circuit 133
Energy Saving Act 159
Engine life 76
Engine management 20, 24
Engine oil 7, 117, 139
Engine speed 8
EPA 156, 164, 165, 167
Ester 117, 134
Ethanol 123, 126, 127, 132, 167
EUDC 148
Evaporation cooling 36
Evaporative emission 175
Excess air 8, 10, 11, 24, 89
Exhaust aftertreatment 7, 11, 20, 61, 86, 132
Exhaust backpressure 90, 91, 106, 110
Exhaust composition 7
Exhaust emission 10, 56, 131
Exhaust emission treatment 106
Exhaust gas recirculation, external 12, 20, 28, 57, 69, 82, 111, 131
Exhaust gas recirculation, internal 20, 28, 29
Exhaust recirculation rate 27, 70
Exhaust section 9
Exhaust temperature 28, 41, 103, 111
External mixture preparation 21, 81

Ferrocen 113
Fiber filter 108
Filter 20, 61, 106, 113, 115, 154
Filter efficiency 106, 110
Filter material 106
Flame ionization detector 152, 153
Flame quenching 21
Flame speed 40
Forest damage 15
Formaldehyde 119, 168
Four-stroke engine 20
Four-valve technology 70, 71
Friction power 78
FTP-72 156
FTP-75 61, 62, 119, 143, 149, 150
Fuel 13, 102, 117, 143
Fuel additive 136
Fuel consumption 2, 20, 28, 32, 35, 37, 41, 49, 51, 55, 58, 71, 77, 78, 80, 98, 102, 112, 117, 118, 142–146, 156, 159, 166, 171, 173, 177
Fuel costs 146
Fuel injection, hydraulic 54–56, 64
Fuel modification 11, 70, 119
Fuel quantity 20, 26, 69
Fuel tax 179

Gas temperature 34
Gasoline 8, 117, 119, 127, 132, 144
Geometric compression ratio 43
Glow pencil control 68
Glow plug position 49, 51
Greenhouse effect 1, 7, 14, 18, 145

Subject index

Haemoglobine 6
HC emission 9, 12, 22, 24, 27, 31, 34, 36, 39, 45, 48, 50, 51, 57, 58, 69, 81, 87, 95, 105, 119, 122, 126–128, 133–135, 138, 164
Head scavenging 78, 79, 82
Heating curve 51
High vaporization 124
Higher final compression temperature 43
Hot cooling 36
Hot starting 149
Hydrocarbon 3, 17, 89
Hydrogen 123, 129
Hydrogen mix operation 131, 132
Hydrogen operation 37, 130
Hydroxyl 17

Idle exhaust test 159
Idle stability 30
Ignition delay 65, 66, 138
Ignition energy 26, 27
Ignition jet method 133
Ignition quality 120–122
Ignition temperature 113
Ignition timing 8, 20, 22, 25–28, 44, 50
Inclined injection 52, 54
Infrared measuring instrument 152
Injecting secondary air 86, 87
Injection angle 53
Injection duration 49–51, 53
Injection jet 24, 52, 63
Injection nozzle 24, 25, 33, 56, 63, 64, 133, 134
Injection port 50
Injection pressure 49, 50, 65–67, 70
Injection process 49
Injection quantity 25, 26, 47, 62
Injection rate control 62, 63
Injection system 24–26, 46, 81–83, 133
Inlet port 32
Inlet valve 26, 58
Insoluble particulate component 12, 105, 106
Inspection and Maintenance program 171
Intake air preheating 111
Intake air restriction/throttling 111, 112
Intercooling 57
Internal combustion process 20
Internal efficiency 34
Internal mixture formation 21, 42, 81

Japan 157, 172
Japan test cycle 173

Knock characteristics 28, 36, 117
Knock limit 22

Lack of homogeneity 12
Lambda control 8, 9, 26, 89, 99, 101
Lambda map 41
Lambda range 42, 89, 101, 128
Lambda sensor 26, 42, 95, 100, 172
Lambda window 96, 98
Latent heat accumulator 20
Laws regulating the emission 147, 156

Lead 117
Lead additive 10
Lead compounds 4, 10
Lead emission 117
Lean burn catalyst 18, 47, 62, 86, 89, 103
Lean burn concept 37
Lean burn engine 36, 38, 39, 47, 102
Lean burn management 39, 41
Lean burn operation characteristics 38
Lean burn operation limit 38
Lean mix engine 40, 41, 47
Lean sensor 42
LEV 94, 168, 171
Light-off performance 91, 94, 104
Light-off temperature 94, 105
Light-running oil 141
Long-term exhaust test 75, 164, 165, 171
Long-term stability 72, 75, 76, 112
Loop-scavenging 77, 78, 82
Low-sulphur fuel 63
Lubricant contents 13
Lubricating film 9

M90 127
Main chamber 50, 51
MAK value 6
Metal-substrate catalytic converter 91, 98
Methane 3, 18
Methanol 123, 124, 126, 127, 132, 144, 165, 167
Misfire limit 8, 9, 24, 37, 124, 125, 130
Mixture control 24
Mixture control system 26
Mixture cooling down 122, 123, 131
Mixture distribution 26
Mixture formation system 44–46, 98
Mixture injection 44–46, 81
Mixture movement 21
Mixture preparation/formation 24, 81, 125, 131
Mixture stratification 33
MON 117
Monolith 90
Motor vehicle traffic 17
Multi-hole nozzle 64
Multi-jet process 48
Multi-point injection 24
Multi-valve engine 24, 28, 59

Natural gas 123, 136
NDIR measuring instrument 152, 153
NH_3 103, 131
Nitrogen compounds 4
Nitrogen dioxide 6
Nitrogen oxide 4, 10, 12, 16, 21, 22, 126, 173
Nitrous oxide 18
NMHC 165, 168, 170
NMOG 168, 170
NO_2 12
NO_x concentration 12, 23, 34, 48
NO_x emission 10, 17, 21–24, 26, 28, 29, 38, 41, 49–51, 54, 56–58, 61–65, 67, 69, 80, 82, 86, 89, 103, 120, 123, 127, 128, 130, 131, 135, 138, 165
NSCR 103

OBD II 166, 171
OBVR 167
Octane rating 138
Oil additive 139, 140
Opening timing valve 32
Operating limit 27
Orbital engine 79, 80
Overrunning mode 99
Oxidant 113
Oxidation 3, 110
Oxidation catalyst 12, 39, 57, 62, 86, 88, 104, 135
Oxygen analyser 100, 153
Oxygen content 9, 10, 101
Oxygen storage capacity 92
Ozone 15, 17, 18, 119, 127, 157, 165, 167
–, chemical processes 17
Ozone depletion 14
Ozone pollution 14, 15, 17

PAK 57
Palladium 57, 92, 105
Paraffin 10
Particulate 4, 6, 12, 47, 53, 54, 57, 62, 67, 89, 105, 107, 109, 119, 122, 132, 138, 139, 172
Particulate composition 13
Particulate measuring instrument 152, 154
Particulate sampling 150, 155
Particulate sizes 106, 107
Particulate trap 61, 62, 110
Passenger vehicle 1, 104, 142
Petroleum gas 123, 128
Phase converter 28
Photochemical reaction 17
Photochemical smog 14
Photooxidants 14, 15
Piston bowl 56
Platinum 92, 100
Poison 6
Pollutant 7, 10, 20
– components 8, 11
– concentration 7
– emission 2, 29, 53, 69
Pollution volume 64
Port control 78
Port cutout 33
Post-oxidation 11, 47
Post-reaction 22, 34, 86
Pre-chamber engine 48, 49, 69
Pre-combustion engine 12, 49, 51, 70
Pre-injection 62, 65
Precious metal coat 92, 93, 98
Preinjection angle 24
Premium gasoline 117
Pressure gradient 51, 66
Pressure loss 92
Pressure-atomizing direct-injection 44, 46
Primary energy consumption 142, 144
Propane 129
Pump nozzle 61, 67

Quality management 11, 41, 43, 44, 81, 124
Quench effect 9

Radiation exchange 18
Rape seed methyl ester 134, 135
Raw emission 7, 38
Reaction 3
Reaction mechanism 4, 5
Reaction of methanol figure 5
Reaction velocity 3
Reactor, thermal 86, 88, 102
Recycling catalytic converter 100
Reducing of fuel consumption 2, 47, 61, 143
Reductant 103
Reduction, catalytic 103
Reduction, non catalytic 103
Reduction process 89, 90, 103
Reformulated diesel fuel 122
Reformulated gasoline 119
Regeneration 106, 110, 111
–, catalytic 113
–, electrical 112
Regeneration burner 106, 110, 112
Regeneration intervals 110
Regular gasoline 117
Residual gas content 28, 80
Retarded idle setting 31
Rhodium 92, 100, 105
RME 134
RON 92, 93
Rough running 41

Sac hole nozzle 65
Sampling 150
Scavenging blower system 79
Scavenging losses 81
SCR 103
Secondary air injection 20, 87
Secondary air map 86
Separation/collector system 106, 107
Sequential injection 25
Service life expectancy filter 113
Shed test 157, 160, 176
Simultaneous injection 26
Single-jet process 48
Sinter metal filter 109
Smog 12
Smoke analyser 154
Smoke number 47, 50, 51, 54, 120
SO_2 12, 105
SO_3 121
Solar energy 130, 145
Soluble particulate component 12, 105, 106
Soot 6, 49, 104
Soot formation process 13, 47
Soot number 50, 51, 64
Spark ignition engine 1, 10, 11, 20
Spark plug 9, 23, 26, 27
Spark plug position 23, 24, 27
Specific consumption 45, 50, 51, 58, 60
Specific surface 96
Specific work 43, 44, 54, 69
Squish areas 9, 21, 24
Start of injection 24, 30, 49, 50, 53, 54, 58, 68
Start opening inlet 30
Start-up catalytic converter 98

Steel wool filter 108
Stoichiometric air quantity 41
Stored emission 137
Stratified charge 40
Stratosphere 14–17
Stroke 34
Substrate 90, 91
Suitability for leaner setting 11, 22, 28, 34, 125
Sulphate formation 12, 13, 57, 104–106, 121
Sulphur contamination 104
Sulphur content fuel 6, 104, 105, 120, 121, 174
Sulphur dioxide 12, 104, 121
Sun radiation 13
Supercharging 50, 54, 56, 60
Surface of the filter 107
Surface-to-volume ratio 21
Surface-type trap 107
Surveillance test 166, 171
Swirl 32, 54, 55, 65
Swirl chamber 12, 49, 73
Swirl chamber engine 44, 48, 50, 52, 56, 73, 120
Swirl inlet port 54
Swirl number 65
Swirl port 24, 55
Synthetic oil 140

Tax formula 179
Temperature increase in earth atmosphere 18
Ten-mode test cycle 159, 160, 173
Test gas mixture 155
Test(ing) method 20, 147, 171
Testing procedures 106, 147, 148
Thermal efficiency 32, 43
Thermal reactor 20
Three-mode test 159
Three-way catalytic converter 8, 12, 18, 26, 39, 80, 86, 89, 90, 98–100, 103
Throttle 32, 44
Tier 2 Standard 165
TLEV 94, 171
Toluene 118
Toxicity 7
Transient conditions 20, 26, 47
Transient operating conditions 20
Trap 160
Trap/filter 106, 113, 114

Trap regeneration 106, 110
Troposphere 15, 16
Tumble 33
Turbulence generation 32, 36
Twin-spring nozzle holder 65
Two-stroke diesel engine 79, 82
Two-stroke engine 77
Two-stroke gasoline engine 80, 81
Two-valve engine 28
Type Approval 174, 175
Type approval test 75

ULEV 47, 94, 168, 171
Unburned hydrocarbons 4, 6, 8, 9, 11, 22, 34, 90, 125
Uniflow scavenging 77, 79, 84
US California 157
US City test 148, 149, 157, 165
US Highway test cycle 165
US test procedures 158
USA-49 States 156

Valve cover orifice nozzle 65
Valve layout 23
Valve lift 31, 32
Valve overlap 28, 31
Valve timing 20, 29, 30, 32
Valve train 78
Valves, number of 23, 34, 62
Vanadium oxide 113
Vapor-lock 117, 118
Variable swirl 55
Variable valve drive 23, 28
Vegetable oil methyl ester 123
Vegetable oils 123, 124, 133
Vehicle speed 143, 148
Volatility rate 171

Wall-adhesion 48, 54
Warming-up phase 22, 27, 47, 86, 95
Washcoat 90, 92, 93, 97, 100
Water-fuel emulsion 63

Zeldovic reaction 4, 10
Zeolite 103
ZEV 170, 171

SpringerEngineering

Hans Peter Lenz

Mixture Formation in Spark-Ignition Engines

In collaboration with
W. Böhme, H. Duelli, G. Fraidl, H. Friedl, B. Geringer,
P. Kohoutek, G. Pachta, E. Pucher, and G. Smetana

1992. 352 figures. XVIII, 400 pages.
Cloth DM 169,–, öS 1185,–
ISBN 3-211-82331-X
Translation of Lenz, Gemischbildung bei Ottomotoren,
Die Verbrennungskraftmaschine, Neue Folge, Band 6

The main point represents mixture formation for spark ignition engines, with regard to the basic principles as well as the design and construction of carburettors, injection systems and intake manifolds. In addition, the appertaining measuring techniques are also discussed.

The basic principles of combustion are discussed to such an extent that is necessary to understand the effects of mixture formation.

The book provides an overall insight on the state ot the art, as well as complete information for the industrial engineer, scientific and university students.

Even after more than 100 years since the invention of the spark ignition engine, research is still intensively undertaken in this field, and great progress has been achieved. In this book approx. 400 literature references from all over the world have been taken into consideration.

P.O.Box 89, A-1201 Wien • New York, NY 10010, 175 Fifth Avenue
Heidelberger Platz 3, D-14197 Berlin • Tokyo 113, 3-13, Hongo 3-chome, Bunkyo-ku

*Springer-Verlag
and the Environment*

WE AT SPRINGER-VERLAG FIRMLY BELIEVE THAT AN international science publisher has a special obligation to the environment, and our corporate policies consistently reflect this conviction.

WE ALSO EXPECT OUR BUSINESS PARTNERS – PRINTERS, paper mills, packaging manufacturers, etc. – to commit themselves to using environmentally friendly materials and production processes.

THE PAPER IN THIS BOOK IS MADE FROM NO-CHLORINE pulp and is acid free, in conformance with international standards for paper permanency.